Will It Sell? ™

How to Determine
If Your Invention
Is <u>Profitably Marketable</u>

(Before Wasting Money on a Patent)

James E. White

D1009893

WARNING: This book may be banned in Texas

The Texas Supreme Court's Unauthorized Practice of Law Committee (UPLC) has been granted the right to ban books and/or software that might help you understand the law. The UPLC has met in secret to investigate other books providing information similar to that in this book.

Will It Sell? How to Determine If Your Invention Is Profitably
Marketable (Before Wasting Money on a Patent)

© Copyright 2000 James E. White
All rights reserved

Excepting for brief quotes, no part of this publication may be
reproduced in any form or by any means without the prior written
permission of the author.

Library of Congress Cataloging In Publication Data

*CIP data was applied for but was NOT provided by the LOC because 1) the author contributed
to the book's publication and 2) the publisher had not published 2 other authors yet. If I had done as
other authors have done and created a fictitious DBA press, these "rules" would probably have
been ignored. The cataloging below is not official <u>but the preassigned card number (PCN) is</u>.*

White, James E. (1952-)
  Will It Sell? How to Determine If Your Invention Is Profitably
  Marketable (Before Wasting Money on a Patent) / James E. White
  p.   cm.   — (first of a proposed Will It Sell? series)
Includes bibliographical and World Wide Web references, tables, charts,
illustrations, and an index/concordance.
  ISBN 0-9676494-0-4 (pbk: ph bal. paper)
  1. Inventions—Marketing. 2. Patents. 3. Patent laws and legislation—United
States—Popular works. 4. New products—Management. 5. New
products—Marketing. 6. Success. I. Title. II. Series. III. Subtitle [How...].
[ask LOC]   2000   Library of Congress Catalog Card Number: 99-97368

Published by James E. White & Associates
Marketing Consultants, 4107 Breakwater Dr, Okemos, MI 48864

www.willitsell.com
(517) 381-1960   james-e-white@willitsell.com

Printed in the United States of America
Jan 2000   10 9 8 7 6 5 4 3 2 1

**ERRORS: While considerable effort has been expended to
eliminate errors they may still exist. The author is solely to blame.
Anyone finding an error is asked to report it to the author via one
of the above contact mechanisms. Error corrections will be posted
to the web site for anyone wishing to see them.**

Dedication

To Gina, my wife, without whose indulgence I wouldn't be able to employ myself.

Contents

Contents

CONTENTS

Contents

Asides

Formulas

Homeworks

Figures

Tables

Miscellaneous

Preface

This book quickly covers how I believe you should approach most invention ideas to evaluate their potential <u>profitability</u> in the marketplace. The emphasis is on FREE and inexpensive steps you can take <u>yourself</u>. The major point is that the marketplace does not reward IDEAS, it rewards SATISFYING THE PEOPLE IN THE MARKETPLACE.

Too many inventors suffer from the notion that the IDEA comes first and then wins approval and success through diligence, hard work, and <u>persuasive</u> marketing **by a big company that will run with their idea**. Almost nothing could be farther from the truth although that does, occasionally, happen.

The PROBLEM comes first. The problem may not even be currently recognized by the vast majority of the future buyers because it is so *status quo*. Inventors should concentrate on inventing things that solve common problems (big markets) and for which the VALUE of the solution TO THE BUYER (as provided NEARLY EXCLUSIVELY by the invention) will be <u>perceived</u> as significantly greater than the COST of the invention to the buyer. Those kinds of inventions will be the big winners.

Almost everybody has ideas from time to time that they think just might be big winners—if only they had the money for a patent. You DON'T need the money for a patent—for the right idea selling the product will supply the money for the patent. For even better ideas (but few and far between), you may very well be able to get someone else to pay the money for getting the patent.

Proper weeding of your idea patch <u>before</u> you invest a lot of time and money (not to mention ego) will nearly always give you a far greater chance of success than the most insightful toiling on the first (or only?) idea you have. In fact, the book points out that proper weeding may narrow you down to only one out of a hundred of your invention ideas. Furthermore, it needn't even be your idea—the electric light bulb was not Thomas Edison's idea, he (and a staff of at least 10 and maybe as many as 60) merely pushed through experiment after experiment till they succeeded in creating a <u>commercially viable</u> light bulb.

CONTENTS

If the methods preached in this book are put into practice by the majority of inventors I fully expect the number of patent applications filed annually in the U.S. to drop by anywhere from 50,000-100,000 over the next 3 years. That will be a drop of 19-39% making the number of filings each year about 150,000 and may well put a fair number of patent attorneys out of work. Even though the total number of patents issued would also drop by those percents I would expect the number of worthwhile patents, the ones that make money, to dramatically increase. Perhaps as much as ten times the current level. That would mean the number of profitable patents would go from about 4,500 a year to maybe 50,000.

On the other hand, what if this book showed thousands (maybe even millions) of people who have invention "ideas" each year the way to get their invention to market safely and without too much expense? What if each inventor that now concentrates on 1 idea that never goes anywhere learned to generate 300 ideas a year of which 3 were successfully taken to market? Maybe we should be demanding the training of more patent attorneys to handle the glut of patent applications as valuable patent filings increase to 350,000 or more a year. How do you get your share of that action?

Millions of people have read *Think and Grow Rich* by Napoleon Hill. If you are not one of them then I highly recommend you get the book and read it. Like that book, this one contains a somewhat unstated truth which you are expected to gain through reading the book and applying its principles. Of the millions who have read Mr. Hill's book, I would wager that fewer than a tenth of one percent (probably far fewer) gleaned the unstated truth and applied it. This book's intent is to give you a clear handle on applying that truth to one particular realm to which it is relevant.

One caution. If facts upset you, or might cause you to see yourself or your peer group in a negative light, you may want to put this book down now and go no farther. If you find a fact in error in this book, my sincerest apologies, let me know what it is so it can be corrected. If you find a fact that you believe maligns you or your peer group, deal with the issues necessary to change the fact—then send me an update. Thank you.

Whether you agree or disagree with the emphasis of the book I sincerely hope you profit from the insights you get from reading the many stories it contains. Enjoy!

Acknowledgments

As with any other significant undertaking, a book, does not just instantly pop out of the head and onto paper. While one person, the author, generally gets most of the credit, many others had a hand in the effort. I would like to thank all of the following people for their efforts and suggestions. All but the editor were totally unpaid and generously contributed their time and knowledge. The sequence these people are listed in has no relevance to their contributions, all are appreciated.

First, of course, is my wife, Dr. Virginia M. Ayres, who commented on several versions as progress was made. My twin brother, Jerry White, and his wife, Nancy White, and my mother, Elwanda White, also looked at and commented on multiple versions. Others who commented and/or suggested resources include Andy Gibbs, Doug Gibbs, Michael S. Neustel, Jeremy Gorman, Jack Lander, Ed Zimmer, Dennis McNeely, Mark Wayne, Bob Ross, and George Morgan.

I am also indebted to the stories, some their own and some just being passed on, told 'round the table at the local inventors club meetings in Lansing Michigan. Particularly communicative were Floyd Wallace, Carl Preston, Stan Lyon, Arron Bishop, Bob Cole, and Don Haynes. A special thanks goes to another member Stephen Funk of Master Design for his engineering expertise and his suggestions to keep my discussions of engineering aspects (and other things) on track. My sincere apologies to anyone who I might have forgotten to mention in this group.

Another thank you is to my editor, Elizabeth Stolarek of Writing Wizards, who is to blame for absolutely nothing in the final version of this book. In fact, it is my (perhaps mistaken) impression that she might have occasionally enjoyed my skewering of the rules of the English language to emphasize a point.

I also owe a debt of gratitude to the Corel WordPerfect programmers whose efforts make a self-publishing effort like this possible. Without their "concordance" indexing capability the page identification component of the index would not have been reasonably doable and the table of contents, reference list, formatting, etc. would have been significant chores. (Just for the record, the firecracker/dynamite exploding on the cover is modified from

CONTENTS

clip-art of theirs but all other illustrations, and the cover design, are my own creations.)

Lastly, I would like to thank anybody that contributed but whom (how pretentious!) I accidentally omitted. The omission was not intentional.

None of the above are in any way responsible for the content of this book or any errors it may contain. The errors are my own and I would be delighted to have them pointed out to me so that corrections can be made should subsequent printings or editions come to pass. Also, I am not an attorney and neither I nor the attorneys that reviewed drafts of this book are to be presumed to be giving you legal advice. What you see here is simply my best general advice given my understanding of things.

Thanks again,

Jim

CHAPTER 1

About This Book

This book provides some useful information for inventors (and prospective inventors) but it is also both an advertisement for and an offer of services. If you wish to use the offered services, you should be aware that they are NOT offered for free. If free services are what you seek, go ahead and read the following material, then go elsewhere. I can guarantee you that, in the long run, you will always pay dearly for your FREE services. The cost may not be directly and immediately out of your pocket. More than likely it will be either the cost (in dollars and time) of pursuing something that was not worth pursuing in the first place or the lost opportunity-cost of never correctly marketing a winning product.

Crass Commercialism

Pretty damn brash, huh! You pay money for a book then discover it's a commercial for me. Actually, most business books (probably in excess of 70%) are commercials for their authors, but most are comparatively subtle in the way they provide sufficient information to contact the author or their firm. The subtlety, of course, makes it possible for the author to "sell" the book to a publisher and instantly get access to production and distribution facilities that would not otherwise be so easily available.

In part, this book shows you how to avoid selling your invention under the same (often poor) terms as those authors. On the other hand, if you find the commercial aspect of this book bothers you well beyond the value you derive from the information, you can always request a refund—I promise not to demand access to your T-shirt drawer to verify the veracity of your commercial sensibilities—but I will demand the book be returned with the $250 "FREE" services certificate in it.

If "crass commercialism" somehow violates your sensibilities, you probably are not cut out to be a successful inventor. You will also probably have trouble saying NO to your own invention even though careful evaluation of your invention via the mechanisms suggested in this book indicate that NO is the only commercially viable answer.

Some sources estimate that you will pour an average of $50,000 into a non-viable invention before you finally call it quits. While that money may be a financial loss to you (who are "holy," and not crassly commercial), rest assured that most of your $50,000 will be a financial gain to others who are crassly commercial. If you follow the steps in this book you may be able to keep the total costs for rejecting your 98 out of 100 non-commercial ideas to under $500. You may also be able to keep your final costs to reach a full GO decision on each of your 2 out of 100 commercial ideas to around $5,000 each. Being crassly commercial is in your best interests!

Inventor's First Book

While the book has been intentionally written to be "the" first book an inventor dips into to get a solid handle on the invention process, it should also serve inventors with more experience. The book covers a lot of territory but primarily sticks to giving readers specific steps they can take and the tools needed to execute the steps. The emphasis is on keeping your costs down until you can be pretty certain (based on objective facts rather than your own opinion) that your invention has real, as opposed to theoretical or hopeful, profit potential.

The book could be much longer than it is if it included more information from the many resources that it directs you to. On the other hand, you can think of it as intentionally not overwhelming you with every little detail. In keeping with the emphasis on taking inexpensive steps, you are strongly encouraged to use the identified resources at your local library, nearby business or university libraries, or over the Internet.

As a "first" book, it is also planned as the first book in a series. Other potential topics include: Financing, Prototyping/Modeling, Design/ Engineering, Marketing, and Advertising. If you would be interested in books similar to this one on those topics please let me know. If you would be interested (and capable) of authoring such a book please let me know that too.

Not About Marketing—It's About Getting There

This book is not intended to replace any of the marketing books out there and, in fact, it will direct you to a few of them WHEN YOU REACH THAT STAGE of the invention process. None of the invention process or invention marketing books that I have examined, in my opinion, emphasize nearly enough that **it is the product, and its <u>value</u> to the consumer,** that must be counted on for the real profit rewards—**not the marketing.** This book starts from the premise that a bit of serious up-front effort to determine if there even is a market for your invention will be far more valuable to you in the long run than spending money to "protect" a commercially worthless idea.

Self-Opinion

You may deduce, while reading this material, that I have a rather high opinion of myself. I'm not so sure that is bad. I expect you to have a (justifiably) high opinion of yourself also. But, the caveat is—you (or I) shouldn't automatically lap that high opinion over onto <u>each</u> of our ideas. I've had some pretty boneheaded ideas (maybe even this book) and you may be subject to having boneheaded ideas too. Be prepared to discard ideas by the bushel (or at least set them aside) and winnow down to just your best ones. Once you really discover how cheap ideas are you will be well on your way to only devoting time <u>and money</u> to those that look like they will be the most profitable among the ones at hand.

I'm basically not a nice guy. I expect inventors to be able to totally defend their inventions. If I make a blistering comment and the inventor has no "heat shields" in the form of independent validation, or a serious examination of their competitors, or whatever, for their invention—and no interest in raising any—I won't be interested either. The book, I hope, shows how to start raising those heat shields. My bias is to be quite blunt because I believe it works <u>on the people that have a chance</u>.

I worked in small business sales for a while and virtually the only thing that succeeded in getting a small business person to sell their business for a reasonable valuation was what the Sunbelt Business Brokers franchiser calls an Uncle Harry Talk. In the Uncle Harry Talk you go over the business financial numbers (income, expenses, profit/loss) the owner gave you (assuming they gave you some) and you keep hammering until the owner sees the numbers the same way a prospective buyer does.

<u>Seven out of about 8 small businesses will never sell</u> because the owner refuses to acknowledge that they are NOT making what they would be worth as a salaried employee in their current position in a successful firm. In other words, the BUSINESS is NOT making a profit and the owner isn't even

earning a "fair" wage. That, by the way, does not necessarily mean the owner is a "failure." They could be far happier and more productive as a small business owner than they would be working under someone else. If they like their tradeoff, they are a success.

You will also notice that I SHOUT from time to time. That is because I am afraid that you might be nearly as thick skulled as I am. I shout just to say, "Hey, listen, this is important. I don't want you to miss it." You might disagree with me or ignore it but I hope you'll at least remember I said it.

Negative Bias

The bias of this book, as you may have already discerned, is often quite negative—but do keep an eye out occasionally for the tongue in cheek. The hope is to discourage forever wannabes while giving real inventors solid guidelines for getting profitably marketable products to market. Various draft readers of this book, whose proofreading and comments I sincerely appreciate, have told me I should figure out how to turn some of the negativity around so I can be more positive.

It is my choice (and perhaps mistake) not to do that, first because it differentiates my book from all the others on the market and second because my

Product vs. Invention

You do not have a product until the first manufacturing run produces a ready-for-sale item.

Prior to that you might have an idea which you further refine into an invention before proceeding to product development.

I have attempted to keep my terminology in this book consistent with the above and you should too. Inventors who claim "sales" before they have a product are generally considered untrustworthy and are often ignored. To be ignored is often the worst fate possible to a wannabe inventor.

talks with inventors lead me to conclude that the two-by-four approach is the only one that has a chance at getting the attention of a few of them long enough to focus on the world outside of developing their invention. Those positive books for inventors are generally selling well but they are obviously NOT successfully getting the whole message across. If they were, there would

James E. White

be far fewer patents and the percent that generate profits for the inventor would be far higher.

It is estimated that only 2% of all patents make money for the inventor—the vast majority of those that do are undoubtedly for employee inventors. Employee inventors often get minimal special compensation since their inventions are, typically by contract, assigned to the employer who paid for the time and other development costs and usually directed the invention process.

The employee inventor normally has no capital at risk and gets a steady paycheck regardless of the success of their efforts. The employer usually bears all the costs of development, production, marketing, etc., and stands to lose all invested funds (including the salary paid to the employee) if the invention fails in the market. This book is not intended for the employee inventor.

The Individual Inventor

The individual inventor, on the other hand, does their inventing on their own time and develops their invention with their own (usually) money. If the invention fails in the marketplace, they pay the price in lost time and money. If the invention is successful, however, the individual inventor stands to gain tremendous financial rewards (and maybe even fame). This book assumes you are that type of individual inventor.

Don Debelak, author of *Bringing Your Product to Market* and a marketer who has been helping inventors bring products to market for more than 10 years, believes that "out of every 500 to 1,000 people who try to introduce a new product, only one succeeds." Hopefully there will be some information in this book that will help you avoid lost time and money from pursuing "neat idea" inventions that have no chance of success in the marketplace.

This book is clearly not for people who only want positive feedback. If you want positive feedback there are many organizations and individuals, from patent attorneys to marketers, who will be happy to keep providing that positive feedback <u>for as long as you keep paying them money</u>. By the end of the book you should know how to avoid that IF YOU WANT TO.

Patent Practitioner

I use "patent attorney" throughout this book and often mean "Patent Practitioner" which could be either a Patent Attorney or a Patent Agent. Both have learned the patent game, passed a test, and have other qualifications for doing patent work. Patent Agents cannot do litigation or provide a legal patent infringement opinion but they can do anything else with regard to patents including provide an opinion on your invention versus existing claims.

Patent Agents often have lower overhead expenses than law firms. Also at large law firms you may meet with the "name" attorneys but the odds are your actual work will be done by a less inexperienced attorney being paid only a minor part of your fee.

As my patent attorney said after reviewing this book for me, "I would never give this out to my clients." She went on to explain that the book is too discouraging and she would have trouble getting enough clients to buy patenting services if they first determined if a patent would have any value. She also suggested that I soften the book considerably so that I too could get more clients to pay me. I'll let you make your own decisions on how to proceed but I have not softened the book.

Inventor or Entrepreneur?

This book is primarily geared for inventors but it could be useful to entrepreneurs interested in growing a business and making a buck out of a new invention. Inventors, on the whole, want to show their technical abilities while entrepreneurs, on the whole, want to demonstrate conquering the market and making a (substantial) profit. I've seen estimates that say only 1 in 500 inventors is capable of becoming an entrepreneur but I would wager that the real ratio is closer to 1 in 2—it's just that inventors usually don't want to do what it takes to be entrepreneurial.

I should also note that the word "entrepreneur" connotes some willingness to take risks while that is almost the farthest thing from an inventor's mind. Inventors almost invariably see their solutions as having absolutely no market risk—their invention solution to a problem they saw is successful technically, therefore sales success should automatically follow [NOT!]. It just doesn't work that way. The true entrepreneur doesn't really care if the invention is

James E. White

a 100% technical success, they just want to win in the marketplace (sometimes, to their disgrace, even by less than ethical means).

"True" entrepreneurs are often "just go for it" types that want to charge ahead, bending the rules or not, and skip as many steps, perhaps even a business plan, as possible. "Success" will justify their approach—maybe! This book is not geared for entrepreneurs of that ilk but one that I highly recommend is *Innovation and Entrepreneurship* by Peter Drucker which contains such gems as the following when discussing why Gillette, Xerox, and General Electric were able to sustain long runs dominating their markets by "giving their customers their money's worth":

> "'But this is nothing but elementary marketing,' most readers will protest, and they are right. It is *nothing* but elementary marketing. To start out with the customer's utility, with what the customer buys, with what the realities of the customer are and what the customer's values are—this is what marketing is all about."

My copy has numerous dog-eared pages; I suggest you get your own (or most any other of Peter's books).

Step Sequence

Be aware, too, that this material is presented in a specific order that I believe will make sense for most inventions that you might create. That does not mean, however, that you must always apply the steps to determining profitability in the order presented. Often it makes sense to think about what might be categorized (for a particular invention) as the "killer question(s)." Answer those questions first because you know that you should probably cease further effort if the result is negative.

If you decide, for example, that selling price in a market with competitive solutions is the "killer" question and your invention will be more expensive to manufacture and distribute than competitive (but slightly inferior) products, then you'll know it is probably best to stop work on the idea. But then again, you might want to ask yourself if there might be a high end market of discerning individuals that want your slightly superior result and are willing to pay substantially more to get it. You can buy ink pens, for example, for less than a dime or for more than $350—and they all meet your writing needs.

For purposes of following the sequence of this book, you should assume that the kind of product I have in mind is a product for a fairly broad subset of the population, is relatively simple from the user's perspective, is not seasonal, and is maybe less than half the size of a 2-slice toaster and probably costs under $60 (perhaps well under). It could be an electronic or a mechanical device. It definitely is not an industrial process or even something as large and complicated as a riding lawn mower. Yet even for some of those type things the basic principles should remain the same. The odds, however, are that developing and marketing an industrial process, or something as large and complicated as a riding lawn mower, will take more money, expertise, and time than an individual inventor can devote to it.

I'm primarily thinking of things that are molded plastic, stamped or bent metal, maybe even cast metal, and electronics—all things for which there are no narrow manufacturer cliques. I assume that the inventor will have the ability to clearly enough visualize the invention so that they can sketch the parts without actually building anything. In particular I expect that to be true for mechanical things; for electronics I would not expect that each resistor, etc., would be known, but the general circuitry should be if the inventor is electronically competent.

Normally I think in terms of the inventor also being the designer/developer as opposed to just being the idea person. The process this book describes should be applicable even if you just have the idea and pay for someone else to execute most other steps.

Not Edison?

Keep in mind that, smart as he was and as simple as the invention may seem, Thomas Alva Edison DID NOT INVENT THE LIGHT BULB. He, and a paid staff of at least 10 employees, did "perfect," through months of trial and error testing, and subsequently patent "an Improvement in Electric Lamps, and in the method of manufacturing the same," to quote from patent number 223,898 granted on January 27, 1880. Mr. Edison's light bulb was simply the first truly commercially viable electric lamp in the United States.

An English gentleman (Joseph Swan) patented a similar light bulb in England a few months before Edison and other work considerably pre-dated that. Edison, when he found he couldn't patent his electric light invention in England, assembled a group of investors that bought the rights to Swan's light bulb.

James E. White

On October 8, 1883, the U.S. Patent Office ruled that Edison's U.S. patent was invalid due to prior art by William Sawyer. The light bulb story provides a short glimpse of what this book says. The "light bulb" IDEA wasn't Edison's, successful development took considerable resources, and Edison's patent was worthless well before it had a chance to expire. Pretty scary?

Rules "Guarantee" Success, Right?

While you might be tempted to see this book as a collection of rules to follow, and then, breaking one, decide to ignore the rest also, DON'T. This book is not a collection of hard and fast rules. Examples abound where success came while the inventor clearly violated the specific steps this book advocates. If you wish, you may take that last sentence as a license to do things any old way or to do them exactly like a friend of yours did when they hit it big.

Please, however, don't be shocked when your invention ultimately fails in the marketplace. On the other hand, just because you execute the steps to the letter and make very justifiable GO decisions at each GO/NO GO decision point, don't imagine you are guaranteed success in the marketplace. The final arbiter of your success (or failure), should you get your invention to market, is the marketplace itself.

The (Ugly?) Origins

This book started as a simple presentation to the Michigan Inventors Club in Lansing, Michigan. It evolved into a book after I started to prepare, based on the presentation, a web "page" to go onto a site with local inventor web pages being set up by one of the club members. My web page got too long-winded for a simple "page" so I evolved it into a book.

The book, as originally planned, was to have 3 sections: Determining Marketability, Marketing, and Advertising. However, several things stopped me from creating those last two sections and encouraged me to make the "first section" a book in its own right. After attending the inventors club meetings for a few more months after my presentation and seeing minimal change in the activities of the members, **I realized that the overwhelming tendency of inventors is to ignore marketability issues**.

The book plan was "shortened" and the current title assigned because I believe there is a need for a "first step" book that doesn't encourage you to move on without doing a reasonably thorough job on determining marketability. Despite shortening the plan, the book still came out a pretty decent length. Besides that, it has already killed far more of my time than I expected (you may have that same experience with your invention).

Inventor Stories

In order to make my point, this material frequently describes invention/ inventor stories as I believe I understood them from the inventors themselves, from those who claim to have talked to the inventor, or from printed sources I have read. I have attempted to locate the (supposed) patents in many cases and/or talk to the inventor if I could get name/contact information.

In cases where I have gotten direct information, I have adjusted this material so that it continues to make the points that I believe are essential to an inventor's success but also points out where my misunderstanding was when I originally heard the story. This should be a warning for any inventor who does not have a clear and cogent story or who cannot properly present it to the marketplace. (More on this issue later.)

I must note that a surprising number of my conversations with inventors go along the following lines:

INVENTOR:	I invented X.
ME:	What is the patent number?
INVENTOR:	Oh, I don't remember that.
ME:	What is your name so I can look it up by that?
INVENTOR:	Oh, my name is not on the patent [I do everything through an attorney/I was working for a company/I gave the idea away].

At this point I assume I'm talking with a "wannabe" who isn't even familiar enough with the patent process to know that:

A) regardless of filing through an attorney or a corporation the inventor's name(s) virtually always goes on the patent application and,

B) "If a person who is not the inventor should apply for a patent, the patent, if it were obtained, would be invalid" to quote the USPTO.

What I, and I suspect most people, will believe is:

James E. White

A) the person had a glimmer of an idea—but they never believed in it enough to pursue it, or

B) the person wants credit for something someone else invented first,

so I move on. Neither I nor anyone else can ever help you market something profitably if you don't have any rights to it in the first place—so please don't bother to tell me or anyone else about it. In the eyes of the world YOU ARE NOT THE INVENTOR (maybe not even AN inventor).

A Patent Means Success, Right?

On the other hand, don't fall into the trap of believing you have to have patented something to be an inventor. You can often be far more successful than your inventor friends who run out of money after spending $5,000 or more to get a (possibly worthless) patent.

This book, which, by the way, is not to be construed as legal advice, spends a fair amount of time talking about patents and the issues that surround them. Many books that talk about patents are wholly about patents or specifically devote a chapter or two to them. To my way of thinking they cover the issues but often encourage you to make patent expense decisions on a one-shot basis too early in the process.

The approach taken here is to suggest that you need to understand the basics and only make cost effective patent decisions at points where they minimize your ultimate loss when your final NO GO or product cancellation decision is reached.

Conversely, decisions on some patent issues can dramatically increase your risk of significant loss while maximizing your potential future gain. In all cases, you alone will have to be responsible for your BUSINESS decisions. I highly recommend you make such decisions only after careful consultation with appropriate advisors.

Internet

If you are not already an active Internet user you should sign up today. You will probably need an extra phone line to really take advantage of it without overly disrupting your current life. You will also definitely need a computer but adequate ones are readily available today for under $1,000 and used bargains abound.

At a minimum, get a 486/66 based machine with 16MB of memory, a 28.8 modem, a Super VGA display of 17" or better, and a driver board capable of 1024 x 768 or better. DO NOT let some ape convince you that VGA or 800 X 600 is adequate—if you do, you will find yourself scrolling every-which-way forever and often be totally lost.

While there are numerous offers to provide you with a FREE computer for accessing the Internet I would recommend against these. To get the FREE computer you will have to sign a long contract and make monthly payments that will often total 2-3 times what a substantially better computer would have cost in the first place. Sure, the computer is FREE, but you pay an inflated rate for Internet access AND have a less than appropriate computer for doing other work on. Or worse, a busy signal when you want to get on the Internet.

The bare minimum, not counting the computer, for Internet access through an ISP (Internet Service Provider) is about $16 per month. That will give you "unlimited" access which is sometimes actually limited to 160 hours. That should be adequate. Cheaper services that offer 20 (or so) hours for $9.99 a month are also available but you will probably discover that once you start using the Internet for research it is very easy to exceed their limits.

There are also some "free" access providers that let you log on with no monthly fee provided you are willing to put up with advertising messages being continually displayed on your screen and perhaps having what you access "watched" and recorded by a computer so the ads can be specifically chosen for you.

Two of these free access providers are NetZero and Alta Vista. To start with NetZero call 1-877-638-3117 and order the start-up CD ROM or visit their site at www.netzero.net. For Alta Vista visit www.altavista.com then click "Free Access" on the site's top menu bar. Both sites' sign-up is fairly painless and inexpensive (under $5) even if you have to order their CD ROMs and both have lots of local phone numbers you can use for access.

When you do get signed up with an ISP, especially one you are paying for, you should also consider finding a free mail service other than the one that comes free from the ISP. That way, if you change your ISP you aren't stuck letting everyone (and listservers, etc.) know you have a new e-mail address.

Also be aware, that to sign up for some things on the Internet you may have to have a "paid for" e-mail address. A couple of places to start looking for free e-mail servers are www.hotmail.com and www.mailcity.com. Of those two I prefer MailCity because it has a lot less hokum in it's agreement.

A new service that became available since I started writing this book is www.alladvantage.com, which promises to pay you up to $20 a month at $.50 per hour for using the Internet. The theory is that advertisers will pay big bucks for Internet ads just like they do for TV commercials. Since there is minimal "media" cost for Internet "broadcasts," the media fees that would normally pay for programming content (which is already there for free on the Internet) can be paid to you if you sign up to receive the ads.

Maybe it will work but somehow I just can't get the mathematics to calculate out to where most Internet users can be paid $20 a month to use the Internet. Remember, the money that is to be paid to you must come from advertisers who will only continue to pay for ads in a media if the customers that the advertiser makes their money from respond to the ads on the media well enough to cover the costs to the advertisers.

Having read all of the above you begin to see how COMPETITION is evolving and changing things. Are you still sure you want to go to the expense and trouble of creating a product and entering the fray?

If you cannot afford Internet access or don't like or want to learn computers, you need to partner with someone who can and does. Create a formal, written agreement about how you'll split the profits from your endeavors. I always recommend 50/50 with some agreement as to how costs will be worked out and how disputes will be settled. The appendixes to this book contain a sample agreement.

If you don't want to set yourself up to access the Internet at home or through a partner you can go to most public libraries these days and access it. You will, of course, have to put up with getting to and from the library and living by the library's rules.

This book provides Internet addresses (URLs, Uniform Resource Locators) for locating information or making contact rather than street addresses and phone numbers. The Internet will typically reduce information gathering and contact time from two weeks to a few minutes or a day or two. The Internet addresses were correct at the time they were copied into this book but there is no guarantee they will be current when you read this.

The convention used in this book is to exclude the "http://" that might be required by your Internet browser to invoke the code that fetches the page that the URL identifies. If the previous sentence appears to be unintelligible gibberish, get a couple of hours of basic instruction on the use of your browser and the arcanities of the Internet and you'll do fine, even if you don't understand the bits and bytes of what is going on to make it all work.

In some cases an ending "/" appears where that is what I captured when looking at the site; it may or may not be necessary and different browsers and/or servers may not care one way or the other. If a URL requires capitalization of specific letters it will be shown that way but most sites these days don't care. ALSO, THERE ARE NO SPACES IN URLs; however, for typesetting convenience, I have allowed URLs to have line-breaks after any "/" or "." (slash or period) character.

All links identified in this book (or their updated versions) are available at www.willitsell.com.

URL's just make it possible for me to convey to you my finds on the Internet. The real power of the Internet, or more specifically, the World Wide Web, is the search engines. You can go to specific search engine sites, but my preference is usually to go to sites that either give me a list of search engines to choose from or that use multiple search engines themselves and return the results to me.

For starting points I'll only describe a few search engine sites here. The first is Ask Jeeves at www.askjeeves.com (or the uncensored www.ask.com) at which you can ask your question in plain English. The second is MatchSite at www.matchsite.com where you just type in keywords and can specifically select the engines that are used by its search.

For a web site that just links to a bunch of search engines try zed.8m.com or just Ask Jeeves, "Where do I find good search engines?" On a frequency of use basis I tend to use Lycos (www.lycos.com), Yahoo! (www.yahoo.com), and Ask Jeeves, but it varies by what I want to find and what I perceive the search strengths of the various engines are.

Most engines' home pages include a link to some help or instructions on how to use them correctly to maximize the quality of your search results. The instructions may be tedious, arcane, and, at first, undecipherable, but they are well worth the effort.

Specific Numeric Examples

Sometimes in this book you will find examples with specific numbers cited. I find books that just say "cost varies" (does that mean $1,000 or

$100,000?) to be fairly useless so I have tried to provide some specifics. The specifics are very likely to not apply exactly to your invention. They are intended to give you an idea of the ballpark the costs will be in. It may be possible to locate more specific numbers relative to your industry but you should not count on it. Do some library research, including asking a librarian or two for help. The work you do to move yourself and your idea along will be the best investment you can make in your invention.

Communication Skills

I'm told that it is often best to write a book for comprehension at the mid high school level. I have no doubt that is true. However, despite some encouragement to write at that level, I have made no careful attempt to do so. If you are capable of comprehension at the college level—whether you have any college education or not—you should be able to understand this book.

If you are not capable of comprehension at that level, and are not capable of communicating at that level, you will find considerable difficulty in getting your ideas across to the business people who you will need to assist you in your endeavor. Get a partner who can communicate your ideas successfully. Business people make their money based on sound business decisions, not on bending over backwards to be nice to you.

Like it or not, people have (and will always have) a bias against unclear communication. That bias is particularly acute in the business world because unclear communication robs business people of the non-recoverable commodity of time. If the (your) communication is unclear, the communication, the messenger, and the idea will all be rejected simply because that is most expedient.

Let me offer a couple of clues on how to evaluate your communications ability. The first is to read this book. If you have no trouble understanding it and don't need to go to the dictionary to look up any words, then you can probably communicate clearly enough. If you need to look up 2 or 3 words (whether you actually do or not), you can probably get by with your current level of communications skills. Average high school graduate skills are all that is necessary.

If you have to reread about one sentence every other paragraph to get its meaning, or look up 5 or more words, it's time to think about getting some help. Fighting an uphill battle with a near certain guarantee of failure is a bad business decision when getting help or partnering are so easy. In business terms: 100% of NOTHING is NOTHING; 50% of $1 million is $500,000—which is the better BUSINESS choice?

Evaluating YOUR Communication Skills

The second way to easily(?) evaluate your communications ability is to write what you believe is a clear and cogent description of your invention, how it is to be manufactured, and, at a high level, what your business plan is. You'll need at least 500 words and preferably about 2000 (with a 12-point, proportionally-spaced type font that works out to a little over 2 pages of text).

Do your writing in a word processing program such as Corel WordPerfect or Microsoft Word with the "automatic error detection and/or correction" features off. Do NOT spend a lot of time on rewrites (you can't "rewrite" when you make a presentation in a business boardroom), but correct anything YOU recognize as a "getting it on paper" typo or grammar error. Then have the program evaluate your writing.

First use the spell checker to help fix your spelling. Keep count of the misspellings. Then use the grammar checker to fix your grammar errors. DO NOT just take the grammar checker's word for what is an error! For example, a few paragraphs back Grammatik requested I replace the "2," "3," and "5" with "two," "three," and "five." Those suggested corrections are NOT due to grammar errors, they are due to editing "industry" style rules. (You'll hear more about industry rules later in the book, but, for now, suffice it to say the editing industry, per *The Chicago Manual of Style*, has 17 pages of rules on spelled out versus figure numbers.)

Also Grammatik suggested, among other errors, that I replace "getting help or partnering are so easy" with "getting help or partnering is so easy." While that might be acceptable, it is NOT correct for what I intended to say. Keep count of the true grammar errors you correct but ignore style or incorrect grammar correction suggestions. Finally, have the grammar checker analyze your readability. Write down the count of the number of words in your text. (See Appendix F for instructions on using word processing tools to do all the above or for instructions on manually computing a "fog index" if the tools are not available.)

Now, to figure out how your communication skills rate, divide the number of spelling errors, if any, by the number of words in your text; then do the same for the number of grammatical errors. If you are within the following ranges your skills are probably fine (with the possible exceptions noted in subsequent paragraphs):

Spelling Errors	0-.01
Grammar Errors	0-.001
Flesch-Kincaid Grade Level	8-14
Sentence Complexity	10-80
Vocabulary Complexity	5-50

Homework

IF you did the above, congratulations, you just completed your first "homework" assignment. Groan, I hate homework! You'll find such homework assignments throughout this book. Some will take only a few minutes to complete but many will take days. The investment you make in doing the homework will pay off in the long run, just like your grade school teachers said.

I didn't diligently study my multiplication, etc., tables in grade school, and I still pay the price in slow speed when mental or hand arithmetic is needed. I do, however, diligently do my homework and study the numbers (with a spreadsheet) before I enter into any marketing agreement with any "this will be a big seller" inventor.

My Communication Skills

I'm not a great speller and I have a broad vocabulary so my spelling errors often run closer to the range .01-.015 before spell checking. Not only do I use the word processor's spell checker, I use the dictionary. The dictionary is useful for finding out if I am using a word correctly to say what I mean AND to determine if a word not in the word processor's spelling dictionary is legitimate.

A word like "mentee" is NOT in the dictionary but it is as valid a word as "mentor" for example. Look up "-ee" (a suffix) in your dictionary. Software spelling checkers are still horrendously poor at accepting legitimate prefixed or suffixed words. (In the last sentence "prefixed" was okayed by my spell checker but not "suffixed.") "Spelling" errors, of course, are not "detectable" when speaking but proNUNceeAtion and grammar errors are.

When I did the Grammatik evaluation of the "Let me offer a couple of clues..." paragraphs by themselves (copied into a separate document), and this whole document, I got the results in the following table. For comparison I have also thrown in an analysis of Abraham Lincoln's *Gettysburg Address* and the *Grimm's Fairy Tales* version of *Snow-white*.

The Flesch-Kincaid Grade Level is simply one method of arriving at the approximate level of education necessary for understanding something. Grade 1 is the first year of grade school at around age 6 and grade 12 is the last year of high school at around age 18. The Sentence Complexity and Vocabulary Complexity scores have a minimum of 0 and a maximum of 100.

Reading Level	Clue...	Whole	Gettysburg	Snow-white
Flesch-Kincaid Grade Level	9.28	12.6	12.9	6.9
Sentence Complexity	59	75	75	56
Vocabulary Complexity	17	29	15	4

Join Them Rather Than Try to Blame Them

Among educated people, which you MUST assume most corporate business people you meet or talk to will be, grammar errors are very poorly tolerated. They may not be able to tell you what a *verb transitive* or a *gerund* is—but they'll know for sure if a sentence don't sound right (see). The mental pause they require to "fix" your sentence so that they can be sure they caught your meaning makes them work harder than should be necessary.

Who wants to work harder than necessary if they don't have to? Certainly not the person who won't learn the language but is willing to waste hours and hours getting rejections because they are not understood. I expect 90 percent or better of my readers to pass the communication skills test with ease; but I've already talked to at least two inventors who read early drafts and can't pass the test—yet, I hope. I've also seen, on Internet forums, a significantly higher percentage of inventors whose language skills will make rejection easier than understanding.

James E. White

If your Vocabulary Complexity goes much over 40, and certainly if it goes over 50, you probably are spouting a tremendous amount of jargon. If your invention is a specialized medical device that may <u>possibly</u> be okay, but NOT if your invention is intended to be a household product.

If you refuse to do the "research" of looking up words you don't understand in a dictionary, you'll probably lack the patience to do the basic research necessary for success. If two pages of text is too much to write about your invention, your chances of success are pretty grim. If you don't know the difference between a misspelled word and one that is not in the word processor's dictionary, get someone on your team who really knows English immediately. If you can't tell the difference between a real grammar error, a style suggestion, and a grammar checker error, again, get someone on your team who really knows English.

For persons outside the U.S. who are reading this with an eye to marketing in the U.S.: when my invention is introduced to a foreign market (e.g., France, Germany, Japan) you can bet I'll have paid people fluent in the language and customs of those markets to do the work.

Preachin' Ain't Stop't Sinnin'

You may find this book preachy (and perhaps pompous), and, at times, not politically correct (like using male pronouns). I prefer to use the generic "they" or "them" for he or she or him or her, but occasionally deviate from that when I believe gender is part of the point or best psychographically categorizes the subject at hand. In some cases I am talking about a specific person or inventor without naming them and therefore the pronoun is correct as used. In any event, be aware that this book often advocates that you acknowledge one of your ideas as a "bad" one <u>from a marketing perspective</u>.

You should not let such a "bad" idea trample your self-esteem. I can do that quite well for you. No, seriously, I heartily respect your self-esteem but I want you to have profits to enjoy too. In the long run, your self-esteem will be better served by generating and discarding 98 "bad" ideas while keeping 2 profitably marketable ones—one of which turns out to be a big winner. If you tried to follow through on the 98 non-profitable ideas and after much expense and trouble only saw them die in the marketplace, where would your self-esteem be?

Dry Wit

While I may be preachy, one of my communications "skills" is occasionally deemed to be "dry wit" or "wry humor." You may have already caught on to that. You may have considered it abrasive or grating or ingratiating. Or you may just have thought of it as "the author's" style.

I first learned of my dry wit near the end of my last year in grade school. One of the things the class officers did was write a "will" for the class. The only thing I remember from the will was that my "dry wit" was willed to some lucky party in the 7th grade. Either it wasn't successfully "willed" away or I nurtured what little I had left of it back to full(?) bloom.

You'll encounter more. My intent is to provide a lighter side to the often painful process of letting go of an idea that just doesn't have the oomph to be commercially viable. Once you let go, getting on to the next idea will be easier.

CHAPTER 2

What Is a Patentable Invention?

Pop quiz! Answer the following question:

At its most basic, a patentable invention is:

____a. a great idea

____b. a solution to a problem

____c. both of the above.

The correct answer is "d. none of the above." Perhaps you feel tricked. Perhaps you are getting slightly steamed. Unless you get the tendency to "accept YOUR answer and YOUR answer alone" under control, you are likely to remain a poor inventor forever. How do I justify "d. none of the above" as the "correct" answer.

If you have ever perused the U.S. Patent and Trademark Office (USPTO) site (www.uspto.gov) and/or the IBM Intellectual Property Network site (www.patents.ibm.com), you will quickly see that many patents are neither great ideas nor solutions to any real problem.

Online Patent Search Progress

Incidentally, in just the few months I've been working on this book the changes in patent access via the Internet have been marvelous. When I started the book I had to request and pay for offline prints of patents since 1976 through third party services; you can now get high quality prints online for free from the USPTO site (when the download process was new I had some transfer problems but it seems to work well now).

The USPTO online searching capabilities have been beefed up considerably also and they have now become my primary patent search location rather than the IBM site even though IBM's site has patents back to 1971. For searching of foreign patents the Pipers Patent and Trademark Attorneys site at www.piperpat.co.nz/patoff.html is a good place to start. You can also go to www.dialog.com, but be aware that this is a pay site and finding what you want and doing the search are quite arcane and esoteric, reflecting Dialog's early entry into the online databases game, their extensive coverage, and their searching power.

> The European Patent Office (EPO, www.epo.co.at for the office and ep.espacenet.com for searching), which has abstracts going back farther than 1971 is mentioned here only for completeness. The searching power is extremely limited and a cross search on the USPTO site will show that their results miss more than they catch. "Toilet seat lifter," for example (which is translated into "toilet AND seat AND lifter in title or abstract") netted only 24 hits at EPO but found 66 hits at the IBM site. Maybe I missed something but I can't recommend the site as a good search tool. Reliance on it is certain to cause you to waste a lot of your money.

Patents Are Granted For...

Patents are simply granted to people who (claim to)

"invent or discover any new and useful process, machine, manufacture, or composition of matter, or any new and useful improvement thereof,"

to quote the essence of the U.S. statute governing patents. Per the Constitution of the United States (Article I, Section 8, item 8) the inventor secures, for a limited time, the exclusive right to their discovery. The Patent Office examiners only verify that the description and claims, AS DESCRIBED by their inventors or patent attorneys, are new, unique, and not obvious to the Patent Office.

The examiners DO NOT verify that an invention works or that it can ever be, or never has been, built. They try only to correctly verify that the invention is patentable and has not been patented [in the U.S.] before—and their results have been overturned on more than one occasion.

This is despite the fact that, per the USPTO:

"The patent law specifies that the subject matter must be 'useful.' The term 'useful' in this connection refers to the condition that the subject matter has a useful purpose and also includes operativeness, that is, a machine which will not operate to perform the intended purpose would not be called useful, and therefore would not be granted a patent."

It must be noted here that "[in the U.S.]" above was bracketed specifically because, with the advent of computerized databases of foreign patents, the USPTO now does some checking for foreign patents and will reject your patent application if they find a foreign patent (even if it's yours) that was published more than one year prior to your U.S. filing. Even if they don't find one (because it predates the computerized database or whatever), your patent, if granted, will be declared invalid if it can be shown that a foreign patent existed more than one year prior to your filing or if there is reason to believe you were aware of the foreign invention, and were not its inventor, at the time you claimed you were the inventor in the U.S.

You may also have noted that there was nothing about "ideas" in the above quotes. That is because the patent office does not grant patents for "mere ideas or suggestions." The rules also require that the patent be explicit enough so that anyone "ordinarily skilled in the art" can apply what they learn in the patent to make the invention and make it work successfully.

I'd like to patent the idea to end war, the idea to end hunger, the idea to end poverty, the idea of an anti-gravity platform, the idea of an anti-gravity belt (the idea of an anti-gravity rug is already taken because I saw it in a comic book more than a year ago). The trouble is, I just don't know how to (or how to describe how to) reduce any of those ideas to practice.

What the USPTO Really Does

What really happens is that the USPTO (with one exception) errs on the side of the claimant when granting a patent. They do not have the resources or the expertise to determine if all claims made are in fact correct. Occasionally, an application will be rejected if the patent examiner believes the invention won't work; but the bottom line is that the objection will be removed if the "inventor" simply asserts, "I made one (or I tried it) and it worked" or "I've (self-) studied X extensively and I believe it will work."

The vast majority of granted invalid claim patents expire without any protest because there is no point in wasting time, energy, and money proving the claimant a fool.

When a new inventor comes up with something similar that does work and that does have commercial potential, then, and only then, is the invalid claim patent likely to be challenged. This can happen either because the USPTO wholly or partially denies the claims in the new patent application, or because the invalid-claims patent's inventor claims infringement. If the invalid-claims patent inventor cannot prove to the court that their "invention" works (or that they have been and still are "diligently pursuing" the development of their invention), their patent will generally be declared invalid.

Yes, the invalid-claim claimant pays twice, once to get the (invalid) patent, and once to defend it while it is being declared invalid. The system works well and even I can't think of a better one. To do things otherwise might mean that any inventor, whether independent like you or part of a huge corporation, could "invent" something without knowing whether it would work or not. Then the U.S. taxpayer would foot the bill for the USPTO to do the R & D to build the invention and test it.

What was that one exception to the USPTO erring on the side of the claimant? If you claim perpetual motion as part of your invention, and assert that it works after your patent is denied, the USPTO will probably require that you provide a working version of your invention for them to examine. Hmm... How many anti-gravity devices have been patented?

Shouldn't the USPTO be more diligent and precise? Well, they could be—if the U.S. Congress didn't keep STEALING the money that inventors (and the firms that support them) pay in fees for the Patent Office to process applications. In the past few years Congress has taken over $100 million this way and it may be over $200 million by the time you read this.

If you, as an individual or small business, pay the fees for a patent application, are granted a patent, and keep it in force for 20 years, you'll pay about $4,000 in fees to the USPTO. A large company would pay about twice that.

You also must realize that U.S. patent examiners (of which there are only somewhere between 2100 and 2500) virtually never spend more than about 80 hours and the average is less than 20 hours on your patent. Patent applications in some classes routinely get less than 10 hours of examination. This is despite the fact that it will probably take close to 2 years for you to get your patent. In fact, patent examiners have strict processing quotas that they are expected to meet. They also get bonuses if they exceed the quotas. I think you get the picture. However, you actually get a hell of a lot of value for your money despite Congress' sticky paws and the quota gun the examiners are under.

Unfortunately, on rare occasions, the issues must be decided by a court and AT THE INVENTOR'S (YOUR) EXPENSE. The vociferousness of the few that fall into this category (aided and abetted by the "news" hungry media) tends to exaggerate what is in reality only a very minor problem.

James E. White

Patentable? Commercial?

An illustrative example of patentability versus commercial viability: A friend I met at a local inventors club meeting said he had a great idea for a toilet seat lifter and was fairly certain that none existed on the market and therefore his invention should be patentable and ultimately successful.

We all know that toilet seats being left up (or down) when they shouldn't be is part of the (legendary?/ongoing?) battle of the sexes. I searched the IBM Intellectual Property Network www.patents.ibm.com (formerly, and probably more aptly, named the IBM Patent Server) looking for the strings "toilet" and "seat" and "lift" in titles only. There were 66 toilet seat lifting inventions that contained those strings in the patent title (this is just since 1971 which is the year the online searchable database begins). There were 4,644 patents that might also be relevant because they contained the word "toilet" in the title.

I also used the U.S. Patent and Trademark Office site www.uspto.gov to find the U.S. Classification for Toilet Seat Lifters and found it was "CLASS 4, BATHS, CLOSETS, SINKS, AND SPITTOONS" subclass "246.1 OPENER OR CLOSER FOR A CLOSET SEAT OR LID." (How to do this search will be described later.) A search for "CCL/4/246*" yielded 157 patents since 1976 in the U.S. that contained the relevant class/subclass identification.

I also did a quick search on the Internet for web pages with the words "toilet" and "seat" and "lifter" in them. There were more than a dozen good solid hits. I have contacted all that I could, told them I was writing this book and that their information would help other inventors like themselves, and asked each to provide me with some info on what it cost them to get their patent, get their invention to market, and about how long it took them to recover their expenses. I promised not to divulge anything but aggregate data.

Only one responded and that was to say that, if I asked, he would be happy to tell me why he invented his toilet seat lifter. All sites except one appear to have gotten their patents in the last couple of years. The only one that I believe is profitable is the one that wasn't shouting about their patent. The firm is old and established and their product sells successfully for about $1,500 in the U.S. while the new, PATENTED toilet seat lifters were selling (oops) offered for $39.95 or less. I'll bet that the new firms will be replaced by other new firms within 5 years.

Not only has the lifting and lowering of the toilet seat been addressed in patents, many patents were expressly for inventions that solved problems with toilet seat lifters. Several dealt just with the problem of preventing the seat from slamming down (or up) if deactivation (or activation) of the lifting mechanism was too abrupt (or vigorous). These are problems my friend probably didn't even know he would encounter because he had not yet done any development.

From the preceding material you should conclude that the question is NOT whether an invention is patentable or not.

CHAPTER 3

What Is a Profitably Marketable Invention?

The real question is:

At its most basic, a PROFITABLY MARKETABLE invention is:

_____a. a great idea

_____b. a solution to a problem

_____c. both of the above

_____d. none of the above.

I even gave you "d." this time—but it is not the correct answer. The correct answer is "b. a solution to a problem" and that is the only correct answer.

I hear an immediate chorus of, "but people will buy anything," as exemplified by Hula Hoops, Rubik's Cubes, Superballs, and even sucker rotators. Those things don't solve any problem—do they? Actually, yes, they do. They solve the problem of boredom. In fact, the entertainment industry is one of the largest in the U.S. since most people in the U.S. do not have to spend most of their time engaged in activities just to stay alive.

There is also another huge industry that catches a large share of the so-called "useless" inventions. That is the vanity (or self-worth) industry. While this industry often has items that specifically belong to it (such as jewelry, in my opinion only, some might argue), many products are significantly embellished by designers seeking to position their basic products (automobiles, for example) to appeal to individuals who wish (whether they will acknowledge it or not) to project a certain image or to feel a certain way about themselves.

Will It Sell?

If your goal is to get your invention patented and then to have someone else produce and sell it (pouring on you huge sums of money for the privilege), what do you think that someone else will ask as their first question? The first question is, "Will it sell?" If it won't sell, then ANY money they invest in your invention will be wasted. (Your time, great genius, patenting effort and expense, and devoted enthusiasm for the invention are almost useless to the producer/seller.)

Well, all right then, you will manufacture it and sell it yourself! Nice stubborn streak you got there. Try not to lose too much money (or self-esteem) in the process. Your first question should be "Will it sell?" also—if it appears that it won't (more on how to do this later), then it is time to move on to your next idea. If you only had the one idea you are most likely to always be a "wannabe" inventor.

Supposing it will sell! Hurray, we've got a winner!!! "Oops, not so fast," your prospective producer/seller says. "What will people be willing to pay for it?" Aaahh..., well... that depends on the perceived value of having the problem solved in general, and the perceived value of your solution in particular. A deck of cards and an electronic video virtual reality 3D game machine might well both solve the same problem, but one will sell for a lot more— REGARDLESS of what it costs to produce it. A full scale giant sized arcade video 3D game will be worth even more—but there will be far fewer buyers.

Well, we can always do the old producer trick of selling high-priced ones to the elite buyers first then lowering the price over time (claiming volume economies, of course) to get more people to buy it while maximizing our profits. Nice idea, but that generally only works when the producer/seller's (and often buyer's) perception is that the item will eventually be a mass market item and that production costs, at both the high end and the eventual low end, will still be far enough below the sales price to provide a decent (or maybe even indecent) profit—not to mention your pile of royalties. More about pricing in the next chapter.

Everyday Ideas

Well, what about a profitably marketable invention not having to be a great idea? Look around you. Really see the products you use every day. Most are mundane and ordinary yet they sell profitably. If they didn't they wouldn't be sold for long. Now look at the engineering of them. See how the engineers solved the problems they encountered during development. You could easily come up with technically better, more innovative solutions in many cases. In many cases, you would back off a step or two and redesign the basics so some of the "solutions" wouldn't be necessary in the first place! Granted, most of our everyday products are not "new" inventions, but evolved products. Even evolved products may have patented parts.

Tear Something Up

Don't tear up anything you don't want to be without, but, for a homework exercise, take a close look at 2 or 3 different staplers from different manufacturers or 2 or 3 single line telephones. If you can take them apart, the guts of telephones (or most things) are often the apparent playground of the "get it done fast whatever way you think of first" approach to detail engineering. That isn't necessarily bad—in fact, it should give you a clue about the non-value of devoting resources to essentially irrelevant parts of your own invention.

On the other hand, you will often see quickly and badly done detail engineering that can't help but make the manufacturing cost higher than necessary (a screw where a press-fit would be more than adequate for example). An engineer friend of mine highly recommends *Product Design for Manufacturing and Assembly*, Boothroyd & Dewhurst, 1994 for a classic work on doing the out-of-sight detail engineering. The book is very cheap, only $165, compared to the thousands of dollars you'll spend making millions of your product.

We have now encountered a fundamental point. Most inventors approach "marketing" from the perspective of "How do I get people to buy my invention?" That is the wrong question. The correct question is "What can I invent that people will buy?" That is a pretty big question, but it is one you should use to filter all your invention ideas.

The basic premise of this chapter is (I hope) clear. **People want solutions to problems** (including boredom) and that is what you should provide with your inventions. A secondary factor you should have noted is that, even if the solution is wanted, **the solution must sell profitably**, i.e., sell at a price above what it costs to make and distribute but below what enough customers are willing to pay for it.

Okay, you haven't given up yet. That idea is still banging around in your head and you are pretty sure it is a solution to a problem. You haven't really answered the questions (Will it sell? At what price?) for your producer/seller, but you insist the answers will still be favorable. You're an inventor, you probably have never worked an invention through to market before, and besides, nobody can really know until you try—right? What are the concrete steps you should take to get your invention on its way?

CHAPTER 4

Steps for Idea Development

The basic steps for invention development proceed from Idea Development through Product Development and on to Market Development. **Most inventors leap right to Product Development** (the next chapter) either on the assumption that their idea is great and unique or because it poses an interesting challenge for them. To avoid wasting a lot of time, you must squelch those reactions immediately and put the <u>idea</u> through the following 4-step test.

This series of steps (as do other parts of this material) hit, in a somewhat different fashion, on some of the material in the book *Stand Alone, Inventor! And Make Money With Your New Product Ideas!* by Robert G. Merrick. Mr. Merrick is an inventor whose sole income for over 25 years has come from his inventions. He has taken his inventions from idea through successful marketing. I highly recommend the book, which can be ordered directly from Lee Publishing (408-738-2200). You might also want to see Bob's web site at <u>www.bobmerrick.com</u>.

The Easy Rules First

Mr. Merrick initially asks you to pass your idea through his "Ten Rules for the Stand-Alone Inventor" and I encourage you to get the book and do so. I will not repeat all ten rules here, but the most important ones relative to marketing are:

2. Think Up Products for Big Markets,

3. Invent Products That Can Be Patent Protected,

5. Design Simple Products That Are User Friendly,

6. Develop Products That Offer Repeat Sales, and

10. Price Your Products to Yield a Good Profit.

Excellent reasons for each of these rules are in the book and I won't belabor them here. As he points out, you don't have to follow all the rules—but the more you follow, the better your chances of success. <u>Once you</u>

accept that your idea has successfully passed his ten-rule test (or at least the ones above), you can proceed with the rest of what I believe is necessary.

Patent Issues

For each step, I have started with a few notes on patenting, and, sometimes, other legal considerations. These notes will be distinguished from other text by the vertical bars to the left and right. I am not an attorney, but I believe what I say is correct. I have adjusted things based on the comments of attorneys (who chose to remain nameless) who did get a chance to review this book prior to final publication.

All discussed patenting activities and methods are by no means required. Some, such as an inventor's notebook and good records relative to 1 and 2 below, will likely only have future value if:

1) you do apply for a patent and an interference action occurs (the patent office discovers a similar patent application in process at the same time yours is),

2) you receive a patent and it is later contested,

3) development takes more than a year, or

4) you also want to get foreign patents for the invention.

The rules are often seemingly detailed and precise yet the reality is they are often murky. If you don't generally understand the patent playing field rules after reading this book or the books suggested in the next sentence, GET HELP. A couple of books from Nolo Press that might help are *Patent It Yourself* by David Pressman and *How to Make Patent Drawings Yourself* by Jack Lo and David Pressman. In general, I find Nolo Press (www.nolo.com) books worthwhile in getting me knowledgeable, but I still leave the work to the pros because I don't want to spend a lot of time getting the details right (or worse, wrong).

I've also read a few self-filed patents and my general opinion is often that the inventor wasted the application fee even though they got the patent. The Nolo Press web site is a good site to visit. It has a large section on "Patent/Copyright/Trademark" which is well worth reading if you just want to look at free information. If you wander around much you will also probably encounter this quote discussing enforcing your patent:

"Bringing a patent infringement action can be tricky, because it is possible for the alleged infringer to defend by proving to the court that the patent is really invalid (most often by showing that the PTO made a mistake in issuing the patent in the first place). In a substantial number of patent infringement cases, the patent is found invalid and the lawsuit dismissed, leaving the patent owner in a worse position than before the lawsuit."

To repeat, get help and get it right or you may regret it down the road. On the other hand, just because a patent attorney wrote your patent application, and got it through the process and granted by the USPTO, does not guarantee that your patent is valid. Another important book from Nolo Press is *Trademark: Legal Care for Your Business & Product Name* by Stephen Elias and Kate McGrath.

Freedom of the (Legal) Press Outlawed in Texas

The Nolo site also has several pages of information on why the title page of this book warns that this book may be banned in Texas. The upshot is some lawyers don't like the idea of individuals doing some of their own legal work or understanding the law and thereby avoiding PAYING lawyers.

Carried to its natural conclusion, legislators ought to make the laws totally in secret, convey them to law enforcement people and attorneys only, and bar all reporters from courtrooms or from accessing court records lest anyone not charged with a violation should learn about the law. Law schools should, of course, be shut down so that we could return to the old system whereby, if you wanted to become a lawyer, you'd have to bribe a friend of the king—and get away with it.

Consider: If only lawyers are able to understand the law, then why, on average, are trial lawyers wrong 50% of the time?

"STOP IF" Issues

Scattered throughout this and the next chapter you will find some "THIN ICE" warnings and toward the end of each "step" in this chapter and the next you will also find a block of "STOP IF" issues. These items will be easy to recognize because of the diamond and octagonal "traffic" signs.

The "STOP IF" issues can also be used as a crude self-evaluation. For each issue simply assign a percentage of from 0-100, depending on how probable you believe that statement is a certain fact relative to your invention. **A 0% is the best score and means that there is no probability that this issue will cause the product to fail. A 100% is the worst score and means this issue is certain doom for the invention in the marketplace.** For example, your assignments might look like the following for:

 IF 1. You find your product already on the market!

 0%—I looked in 4 department stores, 5 specialty stores, 5 catalogs for the product category, and consulted 6 experts in the field, and the product was not found on the market.

 25%—I looked in 3 department stores at the nearest mall, including a Sears, and asked a few of my 19-25 year old friends.

 70%—I asked at the local Big K-Mart and the clerk there hadn't heard of anything like it.

 99%—I specialize in another field, and I haven't done any looking yet, but I've never seen it anywhere.

 100%—The first store I went into told me I'd find it at another store and, sure enough, it was there!

Remember, it's not me or anyone else you have to fool, what you want to do is give honest answers to yourself. **It is your money and time that will be lost if you cheat yourself with irrational answers.** To score your invention based on the <u>percentage chance of failure</u> numbers you provide is as simple as the following process:

1. Start with issue 1 and proceed sequentially through the issues recording your score for each issue as you go.

2. Also keep a running total of your score.

3. At the end of each STEP, look at your total score. If it is over 100% you should either move on to your next idea (often the best, but

most distasteful, option) or take a hard look at the issues, starting with the highest percentage, then TAKE THE STEPS NECESSARY to bring each high percentage down to under 5%.

I would suggest that after any step when you cannot get the total score below a 60% chance of failure you should probably move on to a better idea—your scoring is probably optimistic anyway.

After the last "STOP IF" question (at the end of Chapter 5) there are some suggestions for interpreting your overall score. To try to make it easier for you to remember that 0% is a good score and 100% is a bad score I created the following sequence of ecstatic to dead idea light bulbs:

≈0%   5%   15%   25% 35% 45% 55% 65% 75% 85% 95%   100%

If you find assigning probability of failure due to an issue too counterintuitive, you can invert my suggested scoring and assign probabilities of success. The math, however, is slightly more complicated because you then have to subtract each probability of success from 100 (i.e., reinvert) and keep a running subtraction column of probability of success going from 100% down. By going down to 0% and below you avoid that euphoric moment when you first calculate your probability of success as greater than 100%. For item 3 of the preceding scoring process you would then check to see if your probability of success subtraction had fallen below 0%. If it has, then work on the lowest percent issues first, etc.

It would also probably best if you did not try to score the "STOP IF" issues until you've read the entire book once. Too many factors interact so some other issues may be relevant to the specific one you are evaluating. For example, price, quality, simplicity, and safety are all relevant issues when different consumers make "better" or "worse" judgements about your invention versus competing products.

"THIN ICE" Issues

The "THIN ICE" warnings don't need to be scored but they should play a role in your business decision(s) on how to proceed. They mean (I think) exactly what they sound like they mean. If you are on thin ice the odds are against you—but that doesn't mean that a nimble lightweight like you are guaranteed failure if you disregard the odds. Weigh your options carefully and make a calculated business decision. Cold and wet can kill you if you break through the ice, but not necessarily.

STEP 0—Seek the Alternatives ALREADY on the Market.

Patent consideration for this step: Write a clear description of your idea down in your inventor's notebook and sign and date the page. Your inventor's notebook should be a bound (not loose leaf) notebook with page numbers. U.S. law preferentially grants patents to "first to invent," so the date of your invention may become important in the future. Be aware that most foreign patent laws are "first to file" laws. That does NOT generally mean that someone can just take your invention and patent it. In the countries you'll probably want to deal with, the true "inventor" is still the only one allowed to get a patent.

When you are doing the work described in this step, keep your notebook current. Include the dates, the people you talk to, the stores, catalogs, Internet sites, etc.

Once you have the idea fairly well fixed in your mind, and before you even have to tell anybody about it, you must seek the alternatives already on the market. Failure to do so will result in wasted effort (and embarrassment if not litigation) 90.7% of the time. (Of course I made that number up—but I believe its tolerance is acceptable). I have called this Step 0 because, in my opinion, it is not hit as hard or as clearly as it should be in any other book I've seen. This step is also so rarely done by novice inventors that it gives all inventors the unwanted "head-in-the-clouds fool" reputation.

What's the Problem?

First, clarify in your mind the PROBLEM the invention SOLVES. Now go out into the real world (you rarely can adequately do this in your head while in your arm chair) and look at as many situations as practically possible where this problem is likely to occur. Does your invention SOLVE the problem in better than 99% of the situations you see?

Do not fool yourself by adjusting your definition of the problem to more specific situations than you originally did. If your idea for a solution only works in 90% (or less) of the problem situations, and it is not a technically difficult solution and/or for a high cost problem, I would bet that it is unlikely to catch on in the marketplace.

Does Your Invention SOLVE the Problem?

Would you buy a car (transportation product) that only worked 90% of the time? Not today you wouldn't. But when they were "new" in the early 1900s people did because, for the time, cars were technically difficult.

Would you buy a can opener that only worked 95% of the time? No, you might even throw yours away and get a new one when it starts costing you an extra 30 seconds yet only screws up a measly half percent of the time. Would you buy computer software that only worked 90% of the time? You don't have much choice, unfortunately, because software (even to a whiz) still seems to be technically difficult, but the basic technology for cars was worked out years ago.

If you had a rounded lug nut on your car that your lug wrench just spun ineffectually on, would you buy a "rounded lug nut removal tool" that promised to work? What if you then discovered that the tool couldn't possibly work on your car because the lugs stuck out beyond the end of the lug nut the tool had to contact in order to work? Would you be satisfied or ticked?

The rounded lug nut removal tool is a real life example of an invention that, I now understand, was designed primarily for removing the fancy lug nuts that come with fancy wheels and require a manufacturer-specific lug "socket" tool for removal. The inventor may modify the invention to completely address rounded lug nuts too. Despite my new understanding of the nature of the tool, I will continue to assume that there is such a tool as a single purpose, rounded lug nut removal tool because it makes a good example for points that you should consider.

Don't Pick On MY Invention!

Incidentally, if you get stuck on the idea that you don't like the example I use (perhaps because it's to close to your idea) and therefore are justified in righteously ignoring the point, you won't have too far to look to find the person responsible for your product's failure. People will

---

always pick on your invention—but not near as much after YOU make it a smashing success. Me, I'm always more concerned about the ones who heap flattery on me (I wonder what they want?) than the ones that give me an honest, negative opinion (I learn something new I have to overcome).

True, I could simply go to the patent files and find mostly "safe" inventions to pick on (chastity belts, for example, which are still being patented—a recent one includes microcomputer controls and sensors), but for most people (I think) that would not be as instructive as having "intelligent" ideas held up to the light.

Don't get me wrong, I believe **95 plus percent of all inventors are very intelligent people—it's just that they don't always use that intelligence outside the narrow product development aspect of inventing**. Witness the 2% of patents that make money. This book attempts to guide you to rectify that.

If you think that 2% sounds small, realize that there were 243,062 patent applications filed in 1998. Did you see that many new products on the store shelves? Some sources estimate that a bit more than 15,000 new products hit the shelves each year, 80% are gone by the end of the first year, and 95% are gone by the end of the 5th year. Unfortunately I have found no source that indicates what percentage of the new products are "inventions." Most (probably better than 90%) are probably simply new sizes or new flavors or another soap by another manufacturer type "new" products. In other words, not really new or inventions.

For most inventions, I believe your solution must work 98% of the time or more to be successfully marketable. In the lug nut situation a marketable solution might be a set of devices that do work in 98% of all

---

 Your solution works less than 98% of the time.

---

problem situations, but a device that works 70% (more or less) of the time will probably never succeed unless the costs of the alternative are quite high. If your invention doesn't satisfy the (my) 98% rule above—proceed to the next paragraph BEFORE spending more time with your invention to bring it into compliance.

Go Shopping

Whether you are certain your invention satisfies the 98% rule or you know it doesn't, and before you have disclosed it to anyone, go shopping. Even if you don't believe there are any solutions to the problem your invention addresses, go shopping. Go into likely stores that would carry your type of product, boldly walk up to a clerk, explain to them the problem your invention solves, then ask if they stock something that solves the problem. DO NOT divulge your invention. You might just be surprised (wow!) when they lead you right to your invention! Or perhaps (gulp!) an invention even better than yours. "...Darn—I'm not gonna get rich from my idea." Wrong attitude. It is really time for you to get cracking on your next idea—you don't want to be labeled (at least to yourself) a wannabe forever, do you?

Not to worry—it gets worse. That's right, the best scenario for you is that you find out, while your idea is still just in your head, that your solution, or a better one, is already out there. Your wasted expense in time and money is just that one quick trip to a store.

Any Competing Solutions?

Well, what if the clerk can't show you your product or a better one—maybe they just show you a competing solution or 2 or 3... Buy one of each—or at least of the better ones. That costs money! You can't afford it? How serious are you anyway about inventing a (superior) solution to the problem?

You must buy the competing products because you will need to carefully analyze them before yours is developed or marketed. Why? Because—if your product is to sell successfully in the marketplace—it must be at least equal, if not better, on one or more criteria **RELEVANT TO THE BUYER**. Your technical success in doing something a different way that (perhaps) nobody thought of before, is NOT relevant to the buyer.

How You Are Perceived

Keep in mind (especially if you are not willing to buy and analyze competing products) that anybody you deal with in the course of developing and marketing your invention will make a PERCEPTION JUDGEMENT about <u>you and your idea</u> that will be tuned to YOUR behavior toward your idea. If your behavior (in their judgement—which is ALL that counts) is "give me money" or "my idea is better than any you ever had," then you will likely receive minimal support.

On the other hand, if you appear dedicated to solving the problem—whatever it takes—you will probably get as much help as you want (but

---

 You pursue your solution without serious consideration of alternative solutions.

---

it still may not be the answer(s) you want relative to your invention idea and it may mean one of your helpers must actually get credit for the great idea that is the winning solution). Don't believe me? Which scenario below do you think will have the better outcome?

"You've got money and I know you are a smart investor so why don't you put some money into MY great idea. (I know it's a terrific winner—but I can't disclose it to you just yet.) Just let me know how to get in touch with you and I'll tell you where to send your check and how biiiiiiggg to make it. Your money will get ME well on the way to final development and proving of my idea! Thanks."

VERSUS

"I'm looking for help solving the X problem. I have an idea that might work but I need some assistance. The project needs someone with expertise and a bit of financing but I'll be happy to share the credit and profits."

"Gimme That Hammer!"

Getting back to the shopping trip searching for a way to remove a rounded lug nut. What if the clerk says, "But you don't need anything special to do that. I use Vice-Grips and a BFH." You, being unfamiliar with engineering lingo, say "Huh?" The clerk elaborates "Yeah, I grew up on a farm and we had to deal with rounded lug nuts once or twice a year. For car size lug nuts a pair of ten-inch (preferably curved jaw) locking pliers applied tightly and a Big Fancy Hammer or mallet will get it loose with one or two whacks."

The clerk might also say "or you could use a 3 foot piece of 1 inch outside diameter galvanized pipe (technically known as a BFL, a Big Fancy Lever) to extend the handle of the locking plyers." "Uh," you stumble turning red, "really, my problem is a locking lug nut that has a key that prevents free rotation of the outer part of the nut assembly and the little fingers inside it no longer successfully engage the inner lug nut." Now, of course, you are talking a real special problem situation and a real specialty product—does your invention work on it? (I bet not but you'll adapt it.)

Are you just frantically trying to assuage your own ego? Remember, nobody but you even knows you had an idea yet. The clerk, however, is beginning to suspect you may be more than just flaky.

"Fixing" the Problem

The main point here is to be aware that inventors often keep REFINING THE PROBLEM to fit their invention (and often to exclude competing products) rather than the other way around. The smaller the niche you fit your invention into, the less likely it is that it will be profitable (more on this later). A niche that isn't even perceived to be there by your prospective buyers will be a hard one to profitably fill.

Despite having "refined the problem," however, the vast majority of inventors I have spoken with will always start any discussion of their invention with their original, broad statement of the problem. Any intelligent listener soon backs them into their "refined problem" statement and will conclude "kook, not worth more of my time." It is usually easier to become a financially successfully inventor if you avoid being categorized that way.

Of course, sometimes the opposite problem occurs. You wind up with a device with so many bells and whistles that manufacturing costs become

---

 You refined your problem definition to one that is significantly narrower than your original

---

prohibitive. If the buyer only needs a cost effective solution to part of the problem, or to a special circumstance version of the problem, that very well may be where you should direct your effort (more on how to determine what your prospective customers want later).

Well, what if the clerk couldn't show you, or describe to you, a solution to the problem you posed? Are you done with Step 0, can you move on? No. Try another store and another clerk. If after 5 stores you haven't found "your solution" or a better competing solution, stop for a few days and think about it—you may be approaching the wrong kind of stores.

Toilet seat lifters may not be sold at hardware or plumbing stores—they may be sold at prosthetic devices stores or at stores catering to professionals in the field of your invention. Mr. Merrick makes a strong point (and I agree)—inventing outside your field of expertise can be very risky. If you don't even know that specialty stores exist for your category of product, what chance do you think you have against people who do?

Check Thoroughly

While stores and their clerks are the first things I would try, because I get to interact with people, I would also try specific and general Internet searches if nothing turned up at stores. A specific search might be "tarp holder" (you will undoubtedly have to try a number of guesses) while a general search might be "camping equipment." With the narrow (specific) search you might be very UNlikely to turn up your invention idea or other solutions to your problem just because you couldn't guess the right terms. With the wide (general) search you are likely to hit pages or link pages that you can follow to see if what you want is available at stores or catalog houses that cater to the kind of customer you will eventually be seeking. A minimal Internet search will easily kill 2-3 hours.

If you don't find it online, request catalogs (even if they cost $3 apiece) from a dozen catalog places that cater to your kind of customer. Ask friends and neighbors or the library for possible sources or catalog sources of the solution to the problem. Buy a current copy of *The Catalog of Catalogs: The Complete Mail-Order Directory* by Edward L. Palder and most recently published by Woodbine House in March 1999 or look at it at a nearby library. Over 15,000 catalogs are listed.

A Lycos search for the string (including the quotes) "catalog of catalogs" will turn up numerous relevant sites but the best place to order is probably through one of the online bookstores such as Amazon.com (www. amazon.com) or Barnes & Noble (www.barnesandnoble.com.) You can also find many sites on the internet that claim they provide great lists of catalogs, but I have yet to find one that I thought delivered more performance than promise.

That's Not a Competitor, That's Your Market Channel

Incidentally, it may help to think of stores and catalog houses, etc., as FUTURE sellers of your product rather than approaching them with fear and trembling as prospective idea killers. In other words, if you don't find a better solution than yours, your effort to this point is still not wasted—you've got the research on your distribution channel(s) started.

Even if you don't turn up anything via the stores, the Internet, or catalogs, there is still more to do before deciding the coast is clear. Browse the phone book yellow pages (or maybe the business to business yellow pages book) and ask a few people who might know where you could find a solution to the problem. You do not need to know the people before you contact them—you just need to know that they might be able to point you to the solution.

Most people are flattered to be asked for their expertise in helping to solve a problem, just don't make yourself a burden to them (and NEVER burden them with the confession that you had an idea). It's easy, it's cheap, and it's almost anonymous over the telephone. If you have the name of the "expert" ask for them as soon as the phone is answered. If you don't have the name ask whoever answers "Who around your office is the expert on <your problem>?" Try to get the name of the person directing your call before they transfer you so you can start your conversation with "Hi [Mr/Ms/Dr] <name>, <phone answerer's name> tells me you're the expert with <...problems>. I need some way to... [or something that will...]" etc. Even non-rocket scientists are usually justifiably proud of their knowledge and skills.

How Do You Compare?

If you haven't found any competing solutions you are ready to skip on down to Step 1 (pg 50). If you have found one or more competing solutions you MUST do some analysis to decide if proceeding may still be worthwhile. Sit down with the collection of solutions you bought (and notes on any BFH/BFL or other ready solutions that you can come up with or that might have been suggested to you) and look at them closely while answering the following questions.

Is your solution more complex to manufacture than the competition?

Is it harder to use?

Is it more likely to cause accidental injury to a user?

Does it provide less coverage of all possible problem situations? (Don't overlook the fact that the solution must be available for use in the problem situation. If a BFH is easier to carry and is

---

 Your market is too small to attract the attention of major firms. IS the ice firm enough for you?

---

applicable across a broader range of problems, it is more likely to be <u>acquired</u> and kept in a portable tool box than a BFL or a hundred special situation tools. Of course, a specialist (read "small market") might want the tool and keep it at the workbench.)

Might your product or its manufacturing process have a significant detrimental environmental impact?

If the answer to any of the above (or similar) questions is "Yes" (i.e., your product is more complex, harder to use, etc.), you can resign yourself to being

---

 Your solution is more complex to manufacture than the competition.

---

in a (very tough) specialty market—or you can DISCARD your current idea and try to come up with a simpler, easier, etc., idea to solve the problem—or you can move on to your next problem.

In either of these last two cases start STEP 0 over with the new idea. Being in a specialty market, incidentally, may not be a bad deal since you can often charge premium prices and still have the market pretty much to yourself.

(Major firms are usually looking for markets over $5,000,000 annually, and often over $15 or $20 million, otherwise they can't cover their overhead and make a profit.)

 Your solution is harder to use (and provides no significant benefit to trade that off against).

If being in a narrow specialty market looks like a good straw to grab at to save your idea, keep in mind that the most foolish fool is the fool that fooled themselves. Be careful as you proceed with Step 1 etc.

What a Killer Step

If thorough and realistic execution of Step 0 as described above does NOT kill off 90% of your ideas, you are probably too egotistical to ever be more than a wannabe inventor. (Being as egotistical as I am is okay—being more so is a sin!) Note: at this point you really still don't know if your invention will work in practice or even if it can be manufactured. Those questions are never really answered until the Product Development phase where real time and money are spent.

 1. You find your product already on the market!

2. You find a better product than yours on the market!

3. You find 3 or more products generally no better or worse than yours on the market!

4. Your solution works less than 90% of the time in solving the problem...

 IF AND the solution is not considered technically difficult...

 AND the solution does not have considerable savings over alternatives.

5. You are unwilling to honestly evaluate your invention against competing solutions.

6. Your invention could cause serious injury and you are not willing to accept that responsibility.

7. Your invention or its manufacturing process could significantly degrade the environment.

A few comments on the above "STOP IF" issues, and some hints on scoring your self evaluation:

1. <u>You find your product already on the market!</u> It may not be totally what you had envisioned but it will have the fundamental elements. Does your concept have a unique feature that is currently missing? Do you see different styles or designs but otherwise no real differences? A new functional "feature" may still be patentable as your "invention." See page 34 for some suggestions on scoring this issue. (True story: An inventor told me there are no solutions (except his invention) for incontinence during sleep. One, apparently, just didn't materialize at the bedside when he needed it for his mother.)

2. <u>You find a better product than yours on the market!</u> Cheaper (or less expensive) is USUALLY better to the buyer. Style might be a significant decider but may possibly be only good for a design patent on an existing, preferably public domain, product. Breadth of applicability and quality of result are almost always better within 10-20% of the same price. Remember, **you were supposed to go shopping for SOLUTIONS TO THE PROBLEM your invention solves, not for your invention**. If you only looked for YOUR SOLUTION when you went looking, and didn't find it, now would be a good time to start over. Scoring tips:

0%—After a thorough search no competitive products were found OR my invention equals or betters the competitors on breadth and quality but mine will be (I hope) significantly less expensive.

50%—Mine looks about the same difficulty from a manufacturing perspective as competitors and I'm not too sure what price I can get it made for.

70%—There is a product out there within about 20% above my probable price but with a very good feature that my invention can't have.

99%—A major industry player has several solutions and brands that dominate the industry.

100%—There is a product with more capabilities and features and I'm sure I can't substantially under-price it.

3. <u>You find 3 or more products generally no better or worse than yours on the market!</u> If the field is as crowded as this, especially by larger, established players, you'll need a product with a significant "edge" to overcome your late start.

0%—This is not an issue because there are no competing products.

70%—It's a niche field and I think I can provide considerable added value (e.g., service, support) that will win customers away from competitors.

99%—The field is full of very large players.

100%—You think that there are no competing products yet everyone you mention that to spontaneously busts out laughing.

4. <u>Your solution works less than 90% of the time in solving the problem.</u> You can get away with not ALWAYS solving the problem but only if buyers clearly recognize that the product is at the cutting edge of technology or they

appreciate that significant nuances of the problem situations make a "universal" solution impractical.

 0%—Your solution works 99-100% of the time and its simplicity assures buyers that it will continue to.

 25%—Your solution works 90% of the time and the times it doesn't work it is clear that no completely generic solution is probable.

 25%—Your solution works 90% of the time and it clearly is at the forefront of technology.

 50%—Your solution works 70% of the time and it provides buyers with major savings over competing solutions.

 50%—Your solution works 70% of the time and the times it doesn't work it is clear that no completely generic solution is probable.

 50%—Your solution works 70% of the time and it clearly is at the forefront of technology.

 99%—Your solution works exceptionally well in a special niche but competing (including BFH) solutions generally get the job done in that niche too.

 100%—Your solution works 50% of the time or less and it has no technical difficulty and provides no substantial savings.

5. <u>You are unwilling to honestly evaluate your invention against competing solutions.</u> Duh! Where'd that truck come from? I wouldn't step into the street without first looking both ways but I'll leave you to watch out for yourself.

 0%—The sum of your previous "STOP IF" percents is less than 5 and has been conscientiously determined.

 50%—You believe you can succeed at anything if you just put your mind to it.

100%—Your invention doesn't need to be researched because the world will one day see that it is as definitive as the light bulb.

6. <u>Your invention could cause serious injury and you are not willing to accept that responsibility.</u> Let's face it, nothing is risk free. The bathtub, stairs, and kitchen stove account for an alarming number of injuries and deaths each year. In 99.99% of the cases it wasn't the bathtub, stairs, or kitchen stove that was at fault—but the injuries are still real. People in-line roller skating wearing helmets and knee and elbow pads still manage to break their legs. Scissors and knives are inherently dangerous yet cause relatively few serious accidental injuries. The point is, you'll have to make your own peace here.

 0%—You aren't underestimating the risks of your invention and you plan to provide quality instructions and warnings and carry appropriate liability insurance.

 100%—You're scared-to-death someone will be accidentally hurt by your invention.

7. <u>Your invention or its manufacturing process could significantly degrade the environment.</u> If you suspect there will be environmental damage that violates the law, you need to thoroughly research that up front. If any potential damage is use related but not currently illegal, you may want to evaluate the value of your profits to you and the probability your invention will be mandated (or even shamed) out of existence.

 0%—Manufacturing requires no processes that are not well understood and appropriately "controlled" from an environmental perspective.

 60%—Your invention will result in hundreds of millions of discarded aluminum pull-top tabs (or whatever).

 90% —There are some processes whose impact you are not yet sure of.

Whew, if STEP 0 is so hard, are you sure you still want to proceed with the even harder parts of being a successful inventor?

STEP 1—Ask People and Critically Examine Their Responses.

Disclosure Document

Patent consideration for this step: From the U.S. Patent and Trademark Office site (www.uspto.gov) get a copy of the Disclosure Document rules and form. Write up a clear description with illustrations, if useful, for understanding your invention. Prepare a disclosure document package with one or two copies of your disclosure and a check for $10.00 and send it to the patent office. If you send <u>one</u> copy, the USPTO will send back a copy of the cover page with their assigned Disclosure Document No. affixed. If you send <u>two</u> copies, the USPTO will send back one of the copies with their assigned Disclosure Document No. affixed.

The Disclosure Document you file is nothing more than evidence of the date of invention. It gives you no rights but it does start a 2-year clock ticking. The copy of your Disclosure Document kept at the USPTO will be destroyed at the end of 2 years unless you reference it in a patent application within that time. If your development, etc., take more than the 2 years, do not be overly concerned about the destruction by the USPTO. You should still be keeping good documentation and records of your invention and your progress on it so that the USPTO Disclosure Document is NOT your only, or most significant, evidence of the date of invention. For $25 the USPTO will send you a copy of your Disclosure Document, if you ask before it is destroyed.

"Cheaper" Alternatives

Rather than spend the $10.00 on a USPTO Disclosure Document filing, you can get two of your friends to witness your inventor's notebook or you can get your patent attorney to witness your notebook or other documentation describing the invention. If your attorney is the witness, they will likely charge a fee although some do it for free on the assumption you'll be back for the patent application and <u>they will probably want to make a copy and keep it in their files (if they don't suggest this you probably should)</u>. That insures an <u>almost</u> impartial third party has evidence you didn't change your notebook after it was witnessed.

If your friends are witnesses, they MUST be capable of understanding the invention and should write "Read and understood" on each relevant

---

 If you are operating without any form of verifiable records you risk unnecessarily losing your rights.

---

page of your notebook that they sign and date. In court I think an impartially held copy after signing would have more weight than just friends signatures. This is especially true if your notebook is filled with many pages of nuances over months such that it becomes very easy for a witness under examination in court to become confused.

THE FRIENDS MUST NOT STAND TO GAIN FROM YOUR INVENTION (spouses and boy/girl-friends are clearly out). You pick two friends mostly as minimal insurance against the death, incapacitation, or "tainting" of one. A witness is tainted if they stand to gain or are shown possibly not to have understood. Having your document or notebook page(s) notarized <u>might</u> also be effective, but sending a sealed registered letter to yourself (and leaving it sealed) is definitely NOT effective at establishing a date of invention. <u>Having people sign non-disclosure agreements (see the appendixes for examples) as you talk to them for this step is not a bad idea either but you need to balance practicality with risk. At the very least, record, in your inventor's notebook, the names of the people you talk to and the date and time and place that you talked to them.</u>

All you are building is plausible documentation that a jury or judge will hopefully believe should you ever need to go to court with it.

The "Snicker Test"

You don't need the expense of a fancy Focus Group and you don't need to hire anyone for a Mall Intervention. You can stick your ego..., I mean your idea, out in front of people for free. Ask a few of your friends, particularly ones that should be knowledgeable in the area your invention is to be used in. Sometimes this is called the "snicker test" because your idea is likely as not to generate a snicker. It is up to you to interpret the response.

Did they not understand what you described (highly probable if you don't have a model or a prototype to show them)? Did they not want to offend you with an honest opinion? Might their opinion reflect that of potential buyers? Etc. With drawings, or just a word description, most people will <u>not</u> be able to really evaluate an invention, but there are some steps you can take to help.

An Elevator Speech?

First and foremost is: PREPARE A GOOD ELEVATOR SPEECH. A what? Salesmen use them all the time. It's a 60-second speech that you can use on ALMOST ANYONE that clearly identifies the salient points and makes them <u>understandable</u> to ALMOST ANYONE.

You persist in trying to "win" all your discussions about ~~your invention~~ the product you're developing.

When you step into an elevator with a stranger you have very little time to impart to them what you, or your (future) product, are all about, so you need to be clear and concise and ASSUME NO SPECIAL TECHNICAL OR SUBJECT KNOWLEDGE ON THEIR PART. In other words, skip the industry jargon and use words that won't need to be defined for industry outsiders. If you are lucky enough to be talking to an insider, they'll still understand you.

You should also be aware that while inventors are sometimes held in awe, product developers are often more trusted—so call yourself a product developer rather than an inventor. Calling yourself a product developer rather than an inventor also has two other advantages.

The first is people will typically assume you work for a large company and that you are probably not the originator of the idea. Thus they will feel free to be more honest with you. The second is that it lets you distance yourself emotionally from the invention

Don't forget to state EXACTLY (not broadly) what problem the invention solves. Especially for common problems, don't hesitate to mention the limitations of existing solutions but be sure you are right. If your listeners perceive your "problem" as a non-problem, or your existing solution limitations as hokey, your respondents may do or say anything to get away from the "kook" or to change the subject to one that you are sane about.

Also remember how the scoring system works. Let's say hypothetically that you talk to someone and you make 6 points in favor of your invention and they raise 3 points against. Did you win by 3 points? Hell no, YOU lost by 3 points. And that is exactly the way YOU would score it if it were some

---

 You cannot understandably explain your invention to people who are not experts in the field.

---

other inventor raising 6 points in their favor while you quickly saw 3 big flaws in their invention. Write down those 3 flaws before you forget them. You MUST eventually figure out how to deal with those flaws by either fixing your invention or by providing valid (to the buyer) arguments that counter them.

WARNING: Good elevator speeches are not as easy as they might first appear. It is almost certain that your first one or two tries will get blank stares because you assume too much knowledge on the part of your listener. Not everybody understands volts and amps or even leverage, and many inventors themselves don't even have a clue about the second law of thermodynamics. The rule is simple, revise your speech. Raising your voice or becoming more and more insistent will not increase your listener's comprehension.

Industry Jargon

Keep in mind that most industries have terminology which is "understood" in the industry but which is quite wrong when looked at carefully. For example, when I was introduced to telling time as a wee grade school tyke I was taught, "The big hand is on the hour and the little hand is on the minute." You may remember the litany yourself but do you realize that it is WRONG. It is only correct if the "big" hand means the SHORT fat one, not the LONG skinny one (assuming your clock had clearly distinguishable fat and skinny hands).

The correct litany should be "the short hand is on the hour and the long hand is on the minute" (any other consideration of size, in this case, doesn't matter) because long and short are clear and have no ambiguity. You will find thousands of people to this day who teach telling time on analog (not digital) clocks with the incorrect litany. As you have no doubt discerned, I was severely and permanently psychologically damaged by the revelation that my teacher was just flat wrong. Okay, cute story but not a wrong jargon example.

Want an "industrial" example? How about "righty tighty, lefty loosey." It only makes sense if you KNOW you are talking about right hand threaded items, KNOW that the direction means the direction the TOP of the item moves when tightening or loosening, KNOW the axis of the item is relatively perpendicular to the observer/actor and the ground, and KNOW that the observer/actor is facing the direction the item (nut, screw, whatever) will move when the threads engage.

Now that you KNOW all that, lie flat on your back, slide under the front of your car or under the sink, from your underneath position reach up and apply a wrench to the first handy nut or bolt you see on the front of the engine or the drain pipe, then try to apply the rule "righty tighty, lefty loosey" correctly.

Still not satisfied? Wait till you get into the research for your invention. This is particularly true if your invention involves technology that is outside your current expertise. At one point while working on this book (and on an invention of my own) I came across a perfect example where the industry always says X and means NOT X. Unfortunately it was so obvious that I didn't write it down so I have forgotten what it was. (I just came across a similar example when I started to "stem" some fresh strawberries. I actually "de-stemmed" them of course—and then made shortcake.) From an industry perspective it was easy to see why the convention arose. The preponderance of situations called for NOT X so they just abbreviated it to X. In the rare cases they really wanted to specify X they specifically said POSITIVE X.

A Model Is Worth 1000 Words

A 3 dimensional model or even just a reasonable mockup that looks like your invention, but is not necessarily to scale, will also help if your invention is not easy to visually describe in a sentence or two. If 3 or more out of 10 people don't "get" the invention when described in words, a small, pocket-size model is almost an essential. You can't dig in your briefcase for it, you must be able to show it NOW. If your verbal description of the invention uses an existing product as a point of reference, you need to also be careful that

 You don't ask Focus Groups or Mall Intervention respondents to compare you against the competition.

the existing product is either directly relevant or clearly relevant only with respect to shape or size or whatever, so as not to confuse the listener.

A model, by the way, can often be quickly made with cardboard and tape and/or glue (and often paint or Magic Marker) etc.; it doesn't even have to pretend to work. A model (by my definition) will have most of the parts of the real thing, while a mockup will be somewhat cruder and may only substitute pictures, drawings, or blocks for some parts and leave generally invisible parts out entirely. A prototype (at least by my definition) is the real thing, usually accomplished by hand, and is the one and only existent working (at least under ideal or low stress conditions) embodiment of the current concept at the time. During the actual development phase you may go through multiple models and prototypes—but not yet.

Just a note: If building your model starts turning up all kinds of issues you didn't foresee or it starts you down the path of serious revisions to your problem statement, then proceed cautiously. Do a reasonable job on a first pass at the model but use it to focus your efforts on marketing feedback rather than the (perhaps more interesting) issues of development. **Never forget that better than 90% of all inventions fail in the marketplace, not in the technology.**

Get Expert Opinions

After you've gotten the input of a few friends, perfected your elevator speech, and hopefully toughened up your ol' ego a bit, it's time to approach the experts. The experts are anyone you believe will benefit from your invention. That might be neighborhood kids (talk to them in their parents' presence PLEASE) or house spouses or elevator operators or rocket scientists. Yep, you must figure out who in general you need to talk to, identify a few specific individuals, and figure out how to introduce yourself to them—then do it!

Making cold contacts is often a very big hurdle to get over. But what is the worst that can happen to you? You'll bruise your ego? You'll be thought of as a kook? You'll make a new friend? The country's leading expert will volunteer to take your fantastic idea and run with it giving you full credit and 99% of the profits?!

You can also do your own Mall Intervention. Be sure to get permission from the mall's management or the management of the store your type of customer would enter and that you want to do your "intervention" in. If you have a model to present that is fine but not essential. Just ask. Some people won't want to talk to you but some will. Remember, your objective is not to query the whole world—just enough to have a solid idea of whether people might pay for your problem solution.

A book that gives you some good clues on doing focus groups or personal interviews is *How to Develop Successful New Products* by Jerry Patrick (1997). While this book is geared to product line extensions and marketing positioning primarily rather than truly new inventions, it does have one chapter that I consider well worth the read.

Chapter 9 "Exploration and Expansion" is the best I've seen on what to do and what NOT to do in focus group or personal interview situations where you are trying to get some solid feedback on your idea. The essence is not to ask the participants to vote on a "best" idea, it is to get them to tell you what is wrong with each of the ideas (including your new ones and existing products). It is NOT a good idea to present your "great idea" and 3 or 4 "corn-ball" ideas you thought up for "comparison."

The point is that being the "best" of 5 dumb ideas is NOT a good predictor of success; knowing what the prospective buyer wants but is NOT getting should give you some valuable insights. Use your idea and real, competitive (if possible) products. Ask for the negatives but DO NOT try to argue your way past them—write them down for careful analysis later. When you get to the development step, you will want to develop to meet both the current needs that are being met and those that are not.

Is It Real, or Is It Your Memory?

At a minimum, tape record the focus group session(s) or intervention (video, with sound, is much better) for repeated playback and idea mining later. DO THIS WITH THE TAPE RECORDER OR VIDEO CAMERA OBVIOUSLY VISIBLE. You can explain that the recording is being done <u>only</u> to make sure you get their answers right, or ask for permission if you want to (or your state's laws require it).

In focus groups or interventions always be humble. Remember, The Coca-Cola Company essentially asked thousands of taste testers, "Which of these unlabeled beverages taste best?" then treated that answer as if they had asked,

---

 Everybody, including the experts, has the "wrong" opinion about the greatness of your invention.

"Which of these <u>labeled</u> beverages will you buy, whether you've tasted them or not?" You only get answers to questions you ask—and even that is not certain!

Another book worth mentioning here is *Do-It-Yourself Marketing Research* by George Edward Breen first published in 1977 (the one I read) and recently updated and published in 1998. This is a fairly expensive book (over $40) and is geared toward you as a cog in an existing organization, but it still gives you things that are clearly within your capability to do. If you really can't figure out what questions to ask about your idea, this book has dozens of sample pages of questions. The bottom line is, figure out what you need to find out, then ask questions that will get you the answers.

Where, Really, Is Your Market?

Incidentally, think broadly when deciding who might be interested in your invention. Remember, your expenses here are just the cost of talking to people—you are NOT trying to just expand your market in your head with no prospective buyer feedback. Univac was first on the market with computers—but they saw them as exclusively for scientists and reportedly turned away inquiries from businessmen. IBM didn't turn away the businessmen and quickly realized where the really big market was. Some say that it was ten years before IBM computers could compete "scientifically" with Univac's, but by then IBM owned the computer market and Univac was a rapidly fading also-ran.

What a Gut-Wrenching Step

This is a very tough step to get through correctly but you need to do your best and be honest with yourself. Most people, myself included, have a hard enough time just making a cold contact to talk to someone. The difficulty is significantly multiplied when you are putting one of your ideas out there for people to critique. No matter how they evaluate the idea, they will likely very much respect you and your efforts to evaluate the idea. Don't get you and your idea mixed up in your head as a single entity. **Think positive. If the idea gets killed off, you can make more!**

Also be aware that you should NOT be attempting to drag one of your ideas through this step if it should have been discarded in the previous step. That is an almost certain way to get labeled (and get all inventors labeled) as "kook." Generally, if you believe everyone else is so much stupider than you that you can't trust their opinion, it is probably YOUR opinion that is in error. On the other hand, take the examples of the Scanning Tunneling Microscope (STM) and its brother the Atomic Force Microscope (AFM) which were two truly breakthrough inventions in the scientific world.

While the scientific community strove to create devices using shorter and shorter electromagnetic waves or other esoteric techniques to see more and more detail at the atomic level, these two products' inventors found essentially mechanical ways to capture images at the atomic level by mechanically touching (or nearly touching) the atoms. Many in the atomic level microscope development community thought the inventors were crazy working on these mechanical devices when they were first described, yet the inventors received the Nobel Prize for their inventions only five years later. Their first attempt to publish the results of their efforts was rejected as "not interesting enough," but they knew exactly what they had and persevered. Again, that example is not a straw to grab at. Analyze your respondents' responses and your invention's strengths carefully and honestly.

Idea Theft

If you are afraid of someone stealing your idea in this step you should be aware

| NOTICE: More "Patent" Issues |

that it is highly unlikely since few people would be interested in devoting time and money to the development effort (or even just to the patenting effort). You can go to the trouble and expense of the patenting considerations given at the beginning of this step; however, be aware that filing a USPTO Disclosure Document, getting your inventor's notebook signed by witnesses, getting non-disclosure agreements (often abbreviated NDA) signed, etc., do not "protect" your invention.

The bottom line is NOTHING ever "protects" your invention—not even a patent. A patent, IF you can get one for your invention, merely gives you, <u>mostly at your own trouble and expense</u>, "the right to exclude others from making, using, offering for sale or selling the invention throughout the United States or importing the invention into the United States."

While the U.S. Patent office still recognizes "first to invent" as having priority, priority is only applicable if the invention is diligently pursued

 While gathering potential customer reaction you do something that could constitute "publishing."

from the date of its inception. Most foreign governments use a "first to file" rule. While the U.S. allows the inventor to "make public" the invention up to 1 year before filing a patent application, **most foreign governments require the patent application be filed BEFORE any "making public," anywhere in the world, of the invention.**

For example, if you were, as part of STEP 1, to prepare a brochure on your invention and to pass it out without requiring the signing of a non-disclosure agreement, you would likely be giving up all future rights to obtaining a foreign patent. If you offer to sell one of your products, even at some future date, you likely also lose any future foreign patentability and start a one year clock ticking in the U.S. If you must be paranoid, pay for the assistance of an attorney before starting this step.

Theft of product ideas is actually extremely rare but you should also remember that IDEAS are not protectable intellectual property anyway. The USPTO explicitly states "A patent cannot be obtained upon a mere idea or suggestion."

One "theft" I was told about at our local inventors club meeting illustrates the point (among others) very well. An inventor invented a device for determining the polarity of electrical wires in an automobile. The device as patented used a switch that the user flipped back and forth, causing the indicator bulb to go on or off as appropriate to show positive or negative. The theft "claim" by the inventor was that 2 weeks after a patent was awarded (and published by the USPTO), the inventor found an automatic (no switch) polarity tester in an auto parts store.

First, it is not possible to get a product to market in 2 weeks—especially through the auto parts distribution network—so my belief is that the inventor didn't do STEP 0 and go see what was on the market. Second, the idea of an auto wire polarity tester is not patentable, only an embodiment of that idea is, in this case with a switch. Third, the inventor did not simplify his invention (at least from a use perspective and probably even a manufacturing perspective) as much as his competitor did (probably with two bulbs and/or diodes and/or LEDs and/or resistors and/or other standard components).

Fourth, any voltmeter (or multimeter) will provide the same information to a user with only 2 wires instead of the patented device's 3 (2 long wires needed for known (battery) positive and known (battery) negative and a probe). And fifth, an even simpler device with one long wire to known (battery) positive (or negative), one bulb, and a short probe will do the trick if the user understands that the bulb will not light for a same polarity connection but will for an opposite polarity one. (Also see the Giant Paperclip example elsewhere in this book.)

IDEAS are free by law, they cannot be stolen; that, unfortunately, doesn't prevent nincompoop jurors from occasionally awarding money to nincompoop idea claimants. Is that another straw? An invention, "reduced to practice" (even if only via a description on paper), is no longer an IDEA; it is one (perhaps patentable) embodiment of that idea.

On the other hand, fear of theft has probably kept more than 1000 times as many great ideas locked in basements and garages than have ever been illegally stolen. To repeat a point made earlier: **In business terms: 100% of NOTHING is NOTHING; 50% of $1 million is $500,000—which is the better BUSINESS choice?** In the long run, geniuses like you can have many more winning ideas while the thieves activities get known and they rapidly run out of significant victims.

If you keep a good log of who you talked to, and when and where and what you showed them, and someone steals your idea and gets a patent, you will have some expense and trouble but the odds are that you will be able to get their patent invalidated. If the commercial essence of your IDEA was <u>not patentable</u> and someone you talked to about it (without an oral, or preferably written non-disclosure agreement) gets to market first and makes the big bucks—hey, that's the essence of the competitive, capitalistic reality that made the U.S. the great country you enjoy living in. Of course, you should never consider that person as your friend or ever trust them again.

Make no mistake about it, STEP 1 is embarrassing, ego deflating, may seem to require near superhuman guts and determination. <u>Failure to execute it—and accept the results—is probably the biggest reason for expensive failure.</u> There are two tricks that you might be able to use to help with this step. The first is to ALWAYS treat YOURSELF and your IDEA as separate entities. Your idea is not you. If it dies, you'll still live. When people comment (positively or negatively) on your idea they are NOT commenting about or attacking you.

The second trick is: JUST DO IT. That's what they teach all those annoying telemarketers. While you won't have the advantage of an automatic dialing computer placing the calls so that all you have to do is pick up the line, you should think of your effort to talk to people as if they were already on the line or standing right in front of you. Given that, you'll probably find it's easy to talk to them.

In case you still can't handle asking people, hire someone to do it for you. Many marketing firms do Focus Groups and Mall Interventions. Rather than face this step, most inventors apparently hope that getting a patent (even a worthless one) will enable them to get a licensee to pay them for their idea (see, it's patented, it's unique, the government endorses it [NOT!]) despite their failure to show any progress toward answering the question "Will it sell?"

 **IF** 8. Your respondents generally consider a competitive product superior to yours.

9. Your respondents find more problems with your invention than with competitive solutions...

 AND you have no definitive basis for discounting your invention's problems.

Explanations and scoring notes:

8. <u>Your respondents generally consider a competitive product superior to yours.</u> You have to ask for the information you want to know and the respondents have to have all the "facts." You can't put off dealing with price if it's an important issue. Price is only a minor issue, even with many competitors, in the soft drink industry so The Coca-Cola Company was not concerned with it. They were extremely concerned with COMPARATIVE taste to the exclusion of perceived refreshment and brand identification and thus wound up with the wrong answer on "New COKE."

 0%—In all comparisons your invention won hands down.

 10%—In most comparisons your invention won but there are a couple of minor ones that might cause some problems.

 50%—The respondents noted a couple of features missing across the board and you believe you can supply them with your invention.

 70%—Your invention rated about equal with competitors and/or BFH type solutions.

 99%—Your invention lost on most counts so you're going to go back to the drawing board and fix it.

 100%—Your invention always rated third or worse in all comparisons.

9. <u>Your respondents find more problems with your invention than with competitive solutions.</u> **"The only correct answer is b. a solution to a problem."** People want the value of the result, not your invention. Nobody wants drill bits, everybody wants holes. You probably heard that somewhere before but may not have thought of your invention as a "drill bit." If the model you showed people was Avocado Green and everybody said that wouldn't match their kitchen, you may not have a <u>problem</u> with your invention but you do have a style issue to deal with.

 0%—No functional problems were found with your invention.

 2%—Minor problems were found but your invention still beat out all the other alternatives.

 50%—Neither your invention nor any of the alternatives are problem free and none of the problems are major issues to prospective buyers.

 70%—Your invention has one major problem, but all the alternatives have a significant problem or two also.

 100%—Your invention has more problems than any competing solution.

STEP 2—Get a Professional, Reputable, <u>Marketing</u> Evaluation.

Patent considerations for this step: First, get non-disclosure agreements signed by anyone you present your idea to for evaluation.

Do a Patent Search Yourself

To avoid wasting money on evaluations of already patented inventions you might want to do your own patent search or pay to have one done. See Chapter 13 for a more complete discussion of doing your own patent search. First search the patent info available on the Internet. If you find your invention, you should probably stop and move on to your next idea. If you don't find it online, visit your nearest Patent and Trademark Depository Library and do your own search. The complete list of supporting libraries can be found on the USPTO web site (www.uspto.gov). The librarians will train you to do your own search. It's easy but not foolproof. If you don't find anything you'll have to make a judgement call as to whether to pay for a professional search or proceed on the assumption your search was correct.

Your patent search, and even a professional patent search, does not definitively answer the question "Is it patentable?" **The fact that an invention cannot be found by searching in the USPTO's patents does not mean that the invention is patentable. A complete patentability search must consider all prior art, including earlier products, earlier patents, foreign patents, non-patent literature, and "obviousness."** Unless you pay big bucks, a professional patent search will almost always ignore everything but searching U.S. patents. If your invention turns out to be a multi-million dollar winner, you can rest assured that the big boys will do an exhaustive "prior art" search to try to show your patent is invalid.

Yes, you will eventually need to answer, at some significant level, the question "Is it patentable?" but that is not appropriate at this point for a couple of reasons. The first is that a patent attorney's off-the-cuff (free initial consultation) opinion is likely to be based on an oversimplified description by you, and the obvious answer will be, "No (unless there is something special that I can't think of right now)." With that answer, they also avoid any claim on your part that they sucked you in and charged you money when they knew it was a hopeless situation to start with. On the other hand, the patent attorney's instant reaction might be, "Yes, I can get you a patent on that (after you pay me to find out what existing claims you need to avoid and get you to engineer around them and after I make the claims so narrow the resulting patent is virtually commercially worthless)."

The second reason not to get bogged down in the patent process at this point is that, if you have no hope of recovering your patenting expenses because the product has no chance of selling, let alone profitably, why waste the money? True, you won't be an accredited inventor, with your name on a patent for your idea—but you won't go broke either! Are you an inventor to make money or are you an inventor just to be an inventor?

There are a broad range of programs for evaluating inventions, from FREE(?) hucksters who will always tell you to go ahead (and, for a fee, they can help you), to university programs (see following paragraphs), to very expensive market research firms (often used by large <u>successful</u> corporations, is that a clue?). I recommend you go with a service you have to pay some up front fees to—and which spells out what it will do for your money—otherwise you are likely to end up with a huckster who you will pay to stroke your ego (see the Pitfalls to Avoid chapter).

I strongly encourage you to make this contact with a person or organization that has a MARKETING perspective. Remember, your first question should be, "Will it sell?" Without actual products in hand to sell, of course, no marketer can guarantee that their answer to the question will be correct. (Remember the Edsel, a noted failure predicted to do well, and the Slinky, a smashing success predicted to fail.)

University Evaluations

If there is a nearby university or business college you may want to contact them first to see if they offer any kind of small business assistance programs or marketability analysis. The programs can be as simple as one professor looking for "real world" projects for graduate (or even undergraduate) students taking a marketing course. Such student "project" work is probably best for inventions that are enhancements of existing products or that have easily identified competitors since that is typically the type of analysis those courses teach.

More complex, for a fee, programs are probably a better deal, especially for totally new inventions, because they often use experts in the field as well as trained students. In these types of programs your fee <u>may</u> only pay part of the actual costs of the evaluation, the rest being borne by government grants or "volunteer" evaluators. The volunteers can be either doing it at the expense of their own time or at the expense of a willing employer. Once you have used a few of these programs, however, it becomes very clear that your fee probably easily pays more than the direct cost of the service, but it may not be enough to keep the administrative overhead paid in the fashion to which the administrator wishes to be accustomed. I do not try to compete directly with these programs. Rather, I will gladly use these programs on your behalf as part of an evaluation of your invention.

The following information is correct at the time of this writing but there is no guarantee it will be correct when you read it. One of the bigger programs is the Wisconsin Innovation Service Center (WISC) at the University of Wisconsin-Whitewater, 402 McCutchen Hall, Whitewater, WI 53190. Debra Malewicki is the director of the program and I've found her and her staff to be very responsive. They can be contacted by phone at 414-472-1365, by fax at 414-472-1600, or by e-mail at malewicd@uwwvax.uww.edu or innovate@uwwvax.uww.edu. They also have a web site (www.uww.edu/business/innovate/innovate.htm) at which an electronic version of their evaluation application can be found. The biggest "catch" for their program is the fee of $495. They have served over 5,000 clients since 1980. They will sign a non-disclosure form, will do some patent prior art checking (online only), and usually turn around requests in 30-45 days.

The WISC program does the most thorough checking of any of the programs I have used, but I'm not convinced their fee is always worth it. You can do your own online patent search, collect and look through catalogs, talk to retailers in the field of your invention, etc., yourself. They do not do a systematized evaluation like the following evaluators do, but they do provide an evaluator's opinion (like the others) on whether you should proceed or not and what that next course might be. There is no "success probability" score. They also have a follow-up program for $950 to help look for licensees.

The Center for Entrepreneurship, Hankamer School of Business, Baylor University, Waco, TX 76798 is patterned after WISC to the extent that they have many of the same questions on their invention disclosure form. They only charge $150 and promise an evaluation based on 33 criteria that are used to create a Critical Value Score, an Aggregate Value Score, and an Estimate of Success. They also explicitly provide "no confidential relationship" between you and them but will treat your disclosure "with care." They have a web site from which you can get their application form (hsb.baylor.edu/html/cel/ent/inninvp.htm). This is the cheapest of the services I have tried and also the one I felt gave me the least for my money. Essentially you get one person's opinion and no discussion of what it means. Their 33 questions appear to be taken directly from earlier PIES (see next) evaluations. In other words, they have "borrowed" from WISC and WIN and appear to be collecting $150 for little more than 15 minutes of effort.

There is also the Wal-Mart Innovation Network (WIN), Center for Business & Economic Development, Southwest Missouri State University, 901 South National, Springfield, MO 65804. The phone number is (417) 836-

---

 One or more professional evaluators note significant flaws but still give a GO signal.

5671 and the web site is wal-mart.com/win. You can request their registration package through the site. Three weeks after requesting their package I still had not received it so I requested it again. Their claimed turnaround time on submissions is 6-8 weeks; I got mine back in less, but the informational book, *Innovations, Evaluating Potential New Products*, by Gerald Udell, that was supposed to arrive before the report didn't show up till after I complained.

WIN does what is called the PIES-VIII evaluation for $175. This is the 8th iteration of the Preliminary Innovation Evaluation System (PIES) developed by Dr. Gerald G. Udell (and others) when he was working with a National Science Foundation grant while at the University of Oregon. The current version of the PIES book can, I think, be ordered by itself from the Center. (I originally found it on www.amazon.com but can no longer find it there; however, I have found *How to Assess before You Invest: Pies IV - Preliminary (Self) Evaluation System* by Gerald Udell and Kenneth G. Baker at the Barnes & Noble site (www.barnesandnoble.com) which is, I think, superior to Amazon.com when it comes to searching out-of-print books.)

WIN's purpose is not to predict commercial success but to detect serious technical or commercial flaws in the idea submitted. Your fee buys a copy of the book as well as a full set of responses to their standardized evaluation and a discussion of what the results mean. While I was tremendously disappointed in their care in handling details (many typos, etc., in computer output material that is 5 years old or more, clear errors that they insist are correct when specifically asked about them, and having to explicitly ask for my copy of the book), they are still the ones I am most likely to use again.

WIN also no longer submits inventions they believe have merit directly to Wal-Mart unless the production version and packaging are ready. For a repeat fee they will reevaluate and submit (if applicable) when you get to that production/packaging stage. I still found this the best value for the money.

Be aware also that all three of the above evaluation services (WISC, Baylor, WIN) have a "magic" component that makes it impossible for you to reproduce their bottom line answer yourself even after getting back copies of their reports and materials. They use "personal judgement" based on their "objective" answers to the questions, but give you no clue how a change on one of their criteria might affect the bottom line.

For example, if they gave you a score of "50%" (which you might incorrectly interpret as "chance of success") but their answer to "Patentability?" was "don't know" and you subsequently get a patent search and legal opinion that is "emphatically yes," how far, if at all, does that move you above a 50% score?

The obvious marketing ploy in having the "magic" component to the above university evaluations is that inventors must pay the evaluation service fee to get the judgement score. I'm not sure that is what Congress and other grant agencies had in mind when the funds for development and research on these programs were authorized, but it is what we have to live with for now.

For more evaluators, see the list in Bob Merrick's book or contact your nearest business college or Small Business Administration office (www.sba. gov). With all the university programs you get essentially a cookie cutter approach to marketability evaluation. You send your money and a write-up of your invention and they put it through a predefined set of steps and evaluations and return a standardized report with specifics for your invention filled in.

In addition to, or in place of, a university program, you may want to have a professional marketing firm do some research for you. With these firms you can negotiate exactly what you want researched, but be careful not to try to tell them what findings you will accept.

Marketing Firm Evaluations

HALT! STOP! CAUTION! WARNING! We're talking about real marketing firms here, not firms specializing in "submitting your ideas to industry." See the "Pitfalls to Avoid" chapter (pg. 155) for more on this issue.

At the very least you want your paid-for marketing evaluation to not only answer the "Will it sell?" question, but if it's "Yes," to give you some idea of what price buyers might be willing to pay for it and what size the market might be. Make absolutely certain that the market size estimate is for your specific invention and its competitors, not the market size for the industry your invention fits in (i.e., "garden digging implements" not "the gardening industry" as a whole). The professional results might also discuss competitors and their current market shares, depending on the invention.

If you pay enough for the report you should even get some indication of what the competitors are doing in the way of marketing. How much are they spending on various media? What is their market channel? What percent of sales is pull and what is push? Etc. But always keep in mind that the more restricted THE PROBLEM your invention SOLVES, the less reliable this kind of competitor information can be because there is simply no place to get the data.

A market research firm is most likely to <u>try to err</u> on the conservative side in this kind of analysis. Using a conservative estimate lowers YOUR risk because projections of profitability at the conservative level will justifiably induce you to proceed. If you proceed based on profit projections at the most optimistic level possible, on average, you'll lose your shirt! Of course, you can probably expect a minimal professional marketing study to take $3,000 to $10,000 plus your shirt anyway.

Also remember that in the mid 1940s the expert marketing projections were that 1,000 computers would probably be about all that would be sold by the year 2000. Hey, there's another straw!

Other Firm Evaluations

Often firms run by inventors are interested in helping the inventor community out by doing evaluations for a fee. I cannot tell you how to evaluate such firms but I will tell you that the odds are about 75% that an honest firm will tell you that your invention is a commercially bad idea and should not be pursued. That does not mean that if you are lucky enough to fall into the "proceed cautiously" 25% that you have a near guarantee of success. Aside from my firm covered later in the book I'll mention two that I know about and would trust.

The first is Martec Products International where an evaluation will be done by Mark Wayne for $99. The web site is www.martecmpi.com. Mark has a manufacturing engineering background and products of his own on the market. The second is Andy Gibbs' Professional Invention Assessment for

$600 (or more or less). More info can be found at www.patentcafe. com/service/assess.html. Both will caution you that if you want their suggestions for making your invention better, you may get them as a co-inventor (also called a joint inventor) on your patent. Both will also give you all ownership rights despite their co-inventor status. If you are not willing to share inventorship you can tell them up-front that you do not want design suggestions, just a critique.

The issue of co-inventorship does not cause very many patents to be invalidated but it can. This issue is discussed further later in the book.

Self Evaluations

If you really can't afford an outside, objective opinion you can go to the Marketing 101 (Abridged) chapter and see how to do some basic marketing analysis on your own. You can also locate a copy of the book *Millions from the Mind* by Alan R. Tripp and evaluate your invention yourself based on Tripp's Ten Tips starting on page 8. You can also order Gerald Udell's book *Innovations, Evaluating Potential New Products* (the PIES-VIII book). BE AWARE, HOWEVER, THAT YOUR ANSWERS MAY NOT PROVIDE AN OBJECTIVE OPINION. Tripp's book is a very good one to read for examples of successful inventions and broadly covers everything from idea to the eventual big bucks acquisition of your new company by an industry giant.

There is also a free evaluation form that you can fill out online at www.patent-ideas.com which is the site of Neustel Law Offices, Ltd. Among other things Mr. Neustel, a patent attorney, clearly states in his web pages that you should only patent inventions that are commercially viable. If you select the "Invention Evaluation Form" link it will take you to www. uspatentlaw.com/evaluation.htm where you can work your way through the questions. You will probably find you have to struggle a bit with the first few questions but don't give up. Once you get the hang of it, and if you can answer the questions OBJECTIVELY, you'll get a relatively clear answer. Mr. Neustel is also responsible for the National Inventor Fraud Center at www.inventorfraud.com which is well worth a look.

If you have $69.95 to spend, you can go to Andy Gibbs' site at www.patentcafe.com/kits/assesskit.html and order the "Invention Assessment & CD ROM Program" with which you can evaluate as many of your ideas as you want. The CD contains 6 Excel spreadsheets that provide a bunch of good evaluation questions. The seventh spreadsheet collects all your entered numbers and displays your Graphic Performance Report. Andy's product wasn't "done" when I tried it in that the plan was to convert it into a "C" program with a more robust interface, but he's got it on the market! (Andy also has a good free, downloadable self-evaluation form at www.patentcafe. com/inventors_cafe/free_patent_forms.html.)

If you do decide to buy the program I recommend you do it through one of his multi-item packages, you'll get it for what I think is a more reasonable

---

 You did your own "objective" evaluation via one of the inexpensive routes and got a <u>resounding</u> GO signal.

---

price. He is doing what is sometimes called a "live" market test which means it's capable of returning a profit, or at least significant contribution, while he continues development. Such a test is the rule in the software industry and is not uncommon in other industries. Think about this for your invention. Can you get it to the market without having it absolutely perfected? (More on this issue later.)

Andy's evaluation provides a bunch of good evaluation questions and displays a Graphic Performance Report based on your answers. Of course I plugged in the numbers for a jewelry invention I'm working on. My invention didn't do as well as expected, even though it's ready for test marketing, has minimal negative issues, patentability looks good, and "go ahead" scores were received from all other evaluations. I was under the 65% target score line in all but a few places.

The program and the documentation did not give the kind of "clear" answers that other evaluations did. Andy takes considerable pride in that and even gives you lots of hints on how to improve your graph, but in the end you are pretty much on your own. I would caution against "fiddling" to get the numbers up on Andy's or any other self-evaluation. And I would also caution against worrying a lot about numbers beyond the "Will it sell?" issue until you are fairly certain your self-evaluation on that score is accurate.

If you don't want to go to any other resource than this book, you can go through the "STOP IF" issues in the "Steps for Idea Development" and "Steps for Product Development" chapters. [NOTE: No testing of the "STOP IF," or the percentage scoring decision points (found at the end of the next chapter), has been done yet. YOUR feedback will be invaluable.] Knowing just what you know now about your invention, rate each of the issues as to the probability you should stop. Scores range from 0%, if you are certain the issue poses no problem, to 100%, if the issue is a certain killer.

 10. Your invention is already patented...

 AND the patent has expired...

 OR the patent has not expired and you are not willing to be a licensee.

11. One or more paid evaluators see significant flaws in your invention...

 AND you don't immediately see easy ways to overcome the flaws.

Explanations and scoring suggestions:

10. <u>Your invention is already patented.</u> Or something so close to it is that yours would likely be considered obvious. This happens a lot and the smart inventors find that out by doing their own patent searches. It doesn't mean you have to stop. If there is a market for the invention and no one is serving that market now then it might just be a lucrative one for you.

If someone else's patent has expired then it is perfectly legal for you to exactly copy their product and sell it with only one legal exception. If your product, packaging, etc., would lead the consumer to believe, without close inspection, that they were buying the other guy's product (if it's on the market), you may be guilty of fraud or just deceptive trade practices. On the other hand, if the invention is patented and the patent has not expired you (and the patent holder) will need to decide if a licensing agreement is desirable. (Hint: if not, you don't proceed.)

 1%—No conflicting patent was found by both your and a professional search.

5%—No conflicting patent was found in your own search.

 15%—An expired but really close patent was found and the invention is not on the market AND people indicated they liked it in the previous step.

 50%—An expired patent was found and a small company has the product on the market but you think you can do it better.

 50%—More than 5 similar patents were found and the invention is not on the market.

 70%—An active patent was found and you believe that with a license and your new features you could be successful.

 75%—Dozens of similar patents were found and the invention is not on the market.

 99%—Hundreds of similar patents were found and the invention is not sold in any of the major chain stores you would expect to carry it.

 100%—An active patent was found and you are not willing to be a licensee.

11. <u>One or more paid evaluators see significant flaws in your invention.</u> These may vary from environmental to manufacturing to lack of consumer interest. If the expert's opinion is that you should stop working on this idea you had better be able to see a quick and easy solution. Otherwise, you'll likely see expenses skyrocket with little chance of return. Coming up with another idea could be a faster, cheaper, solution for you.

 0%—No evaluator saw any significant problems.

 25%—One evaluator saw one problem but you were able to figure out a solution that will be acceptable to buyers and/or the government and/or your manufacturer within 20 minutes.

90%—An evaluator saw a significant problem and you haven't got a solution but you know of an expert that you think can solve it.

 99%—Multiple significant problems were found and no solutions were found in 2 or more hours of trying.

STEP 3—What Will It Cost to Produce (Approximately)?

> Patent considerations for this step: Get non-disclosure agreements signed by anyone you present your product design or parts to for estimates.

Yes, you must get a solid idea of what it will cost to produce your product before you do final product development but it is not as hard as it looks. First, assume you will NOT be buying equipment or building a factory to make the product. <u>You will be subcontracting to have an existing manufacturer make your product.</u>

Of course that will cut your profit margin significantly. But it will dramatically reduce your start-up costs and will eliminate significant risk.

 You want to build your own factory to manufacture your invention.

It will also save you a lot of time and possibly provide you with a terrific source of expertise that you probably don't already have. If you do not understand this please read Mr. Merrick's book. For his $40,000 in start-up costs for his Personal Punch (business card punch to notch business cards to fit Rolodex files) could he have built a factory, re-invented Dow Chemical's fiberglass reinforced nylon, made injection molds, bought injection equipment, and hired manufacturing staff? Hell no. He might be lucky to do that for less than $4,000,000. Would his punch be profitable—probably not—but he'd have kept more contribution margin and a bigger percent of the sales price.

Margins, Profit & Contribution

Briefly, some "margin" definitions. Don't get too stuck on exact meanings because the terms are often used loosely and interchangeably. Whether the margin is expressed in dollars or as a percent is also irrelevant as long as a percent is compared to a percent and dollars to dollars. In other words, you must understand margins in the context in which they are used. If you don't understand it, and can't get it clarified, discount the statement to some extent.

In a retail setting where no commissions or other direct expenses on item sales occur, "profit margin" is the amount contributed to profit by the selling of an item. <u>After all other expenses are met</u>, the profit margin on the sale of an item is equal to the price paid to the seller for the item (ignoring taxes) minus the cost of the item to the seller.

$$\begin{matrix} \text{Profit} \\ \text{Margin} \\ \text{per Item} \end{matrix} = \left(\begin{matrix} \text{Seller} \\ \text{Receipts} \\ \text{per} \\ \text{Item} \end{matrix} - \begin{matrix} \text{Seller} \\ \text{Cost} \\ \text{per} \\ \text{Item} \end{matrix} \right) - \frac{\text{All Other Expenses}}{\text{Number of Items Sold}}$$

The "contribution margin" is always the price paid for the item to the seller (ignoring taxes) minus the cost of the item to the seller.

$$\begin{matrix} \text{Contribution} \\ \text{Margin} \\ \text{per Item} \end{matrix} = \left(\begin{matrix} \text{Seller} \\ \text{Receipts} \\ \text{per} \\ \text{Item} \end{matrix} - \begin{matrix} \text{Seller} \\ \text{Cost} \\ \text{per} \\ \text{Item} \end{matrix} \right)$$

Obviously the exact "contribution margin" can be determined at the time of sale but the average profit margin cannot be determined until the end of an accounting period when "all other expenses" and the "number of items sold" are known.

In a manufacturing setting the cost of an item is not as easily determined as at the retail level, but it can be generally agreed to be the sum of all direct costs (such as setup, machine operator and assembly labor, and raw materials) attributable to a production run for many copies of an item divided by the number of items produced in the run.

$$\text{Direct Manufacturing Cost per Item} = \frac{\Sigma \text{ A Production Run's Direct Costs}}{\text{Number of Items Made In Run}}$$

Profit margin on a particular product is generally an abstract number to a manufacturer (especially when manufacturing their own products) since the calculation requires many, more or less arbitrary, allocations of "all other expenses" also known as <u>indirect expenses</u> (such as telephone, heat, lighting) to various products or production runs.

Contribution margin, however, is watched closely. The contribution margin of an item is the money received for a specific production run minus all direct costs for the production run divided by the number of items in the run.

$$\text{Manufacturer Contribution Margin per Item} = \frac{\text{Receipts for Run} - \text{Direct Mfg Cost for Run}}{\text{Number of Items in Run}}$$

You could do a "contribution" calculation for all production runs in a year or even for a defined batch in the middle of a production run and those numbers might also give you valuable information for future decisions.

The point is, if you create your own manufacturing facilities and do your own manufacturing those "all other expenses" are probably going to kill your profitability. This will be true even if you have a 2000% contribution margin (you sell for 20 times your direct item production cost) on each of the few hundred items you sell in your first year(s) of business.

On the other hand, subcontracting the manufacturing may cut your contribution margin to a "measly" 20% yet give you a substantial profit because you have only $5,000 in "all other expenses." Even if you just set up in your garage, you need to be real skeptical about the value of buying or leasing equipment.

Contribution Margin Calculation

Do the math on the above paragraph and, at least crudely, for your own invention, as a worthwhile homework exercise. I suspect more than a few inventors have driven themselves to poverty by trying to maximize the percent of the retail price that comes to them rather than concentrating on how much total money ends up in their pile after all their expenses are met.

"Costs" from the Store Shelf

Anyway, back to how to determine your manufacturing cost. First, find a bunch of products that already exist on store shelves that are similar in content and complexity to yours. (You don't need to buy them but try to dress as a respectable person the day you go looking.) These products do not need to be products that compete with yours. They do not even have to be for use in the same field. They need only be comparable from a manufacturing standpoint. Note their retail prices. If they have similar prices use them all. If some have significantly higher prices, ask yourself if they are priced higher due to (perhaps patented) uniqueness or to the cost of the problem they solve. If some are significantly lower, ask yourself if there are many competitors with similar or cheaper solutions.

Recheck to be sure all products are similar in content (plastic parts, metal parts, machined surfaces, etc.) and complexity (number of parts, shapes of parts, etc.) to yours. Exclude the ones that have external reasons for high or low prices and the ones that closer examination shows to not be as similar to yours "manufacturing wise" as they first appeared. Hopefully you have at least 5 or so products left. Sum their prices and divide by the number of products to get the average price. (If your invention is not clearly unique or a solution to a costly problem, this average price might also be about the "comparison price" a reasonable individual buying your product would assume it should sell for—so keep that in mind.)

Now divide your number by 10. Assume that is approximately the DIRECT MANUFACTURING cost of producing the items. Man, are you being ripped off by those greedy company people! Wrong. Remember that

---

 Your first product is quite complex and you don't have the financial depth to place large enough orders.

---

the manufacturer must also cover overhead and make a profit, the distributors must also, and so must the retailer that is watching you ogle their merchandise without any evidence that you might buy something! (More on the profits issue in the "Pitfalls to Avoid" chapter.) So now you know about what it will cost your subcontractor to manufacture each item when thousands are being manufactured.

How much will a subcontractor manufacturer charge you? When they are cranking out thousands it will probably be about 2 times their cost. On the shorter runs that you order when you first start selling it will probably be 3 to 10 times their direct cost. The more complex the product the higher the multiplier for short runs. Make your best guess and be realistic about it. Being overly optimistic (or cautiously pessimistic?) may make the difference between ruin and success further down the road.

Also keep in mind the simple (at most half the size of a two-slice toaster) type product I'm assuming here. For an automobile type item (i.e., a whole car) division by 5 with a short-run multiplier of 100 might be better but still optimistic. If you do the math you'll see why sticking with small, simple items is best for your first few inventions.

Should You Bother a Manufacturer?

So far it hasn't cost you anything (but your mug is on retained surveillance tapes in 3 stores now) to figure out the possible cost to you as the manufacturer of record. But is your guess on manufacturing costs even close?

You're Liable

Incidentally, your subcontract-manufacturer will not be THE manufacturer when it comes to liability or dealing with the distribution channels, YOU will be.

IF your cost still looks acceptable relative to your competitors, now MIGHT be the time to get some expert estimates. If your cost does not look acceptable, it is probably time to go on to your next idea.

If your cost does not look acceptable and you proceed with the next few paragraphs anyway to get some "real" manufacturing prices, you should be aware that you are doing a grave disservice to all your fellow individual inventors out there. Manufacturers "waste" enough time submitting unsuccessful bids for producing goods for companies that do get their product manufactured. Submitting bids on inventions, perhaps yours, that NEVER get manufactured is really a waste of money and effort.

In other words, you're at a small fork in the road. You need to make a judgement call whether your analysis of existing products to arrive at a possible cost is sufficient or whether you need to actually get some bids from manufacturers. I would suggest you get quotes only if all three of the following are met:

1) your invention is fairly simple (i.e., easy and cheap to estimate),

2) there are no material or manufacturability "kinks" you foresee possibly having to be worked out during development, and

3) you are <u>not very confident</u> of your "similar products" cost analysis.

If your invention is quite complex, you foresee some potential "gotchas," and/or you have absolute confidence in your cost analysis, then I suggest that you hold off on getting manufacturing quotes until later when your design is near final and you have good prototypes or models.

Finding a Manufacturer

If you do decide to go ahead and get quotes, there are several avenues of approach:

A. Look in the yellow pages of the nearest big city or the business-to-business yellow pages. To get a business-to-business yellow pages (if there is one for your area) usually only requires a phone call to your phone company and agreeing to pay (about $20) for the book which will be delivered in about a week. This is often the least productive place to start

---

 You want manufacturer quotes on a complex product you don't have engineering drawings and a debugged prototype for.

---

looking since you may have to make a lot of guesses at appropriate headings. Try "Springs" or "Plastics" or "Steel" or whatever might be the key components of your invention. "Machine Shops" might also have some relevant listings for you.

"Prototypes" is probably not a good place to look because you probably won't get manufacturing production pricing. Unfortunately, the business-to-business yellow pages do not contain nearly enough ads to be helpful. Almost all you get are headings and listings. A category, a name, and a phone number gives you the option of calling a company and asking what they really do, etc., but you have to do a lot of work to narrow down your choices.

B. Look in *Thomas' Register* at any decent sized library or online at www.thomasregister.com. Online you'll have to register first but it's free; the real problem will be you'll have another new ID and password to remember. You'll want to ignore listings for your direct competition and go for more generic manufacturing concerns that appear to have the capabilities needed for your type of product.

C. Try *Harris Info Online* at www.harrisinfoonline.com. At this site you can rummage around without registering first. You need to be aware, however, that the coverage seems to me to be considerably poorer than *Thomas'* even though a claim on Manufacturers Info Net says they are "the most accurate, in-depth and comprehensive database of American Manufacturers available." Also, if you want more than just basic contact info you will have to pay for it.

D. Start rummaging from one of these sites: Manufacturers Info Net at mfginfo.com or Industry.Net at www.industry.net. Or do your own search via one of the major search engines. If you don't have a computer and access to the World Wide Web, then either get it or be willing to pay someone to do Internet research for you. If you don't use the Internet, you'll almost always find yourself eating dust these days.

From any of the B through D options above you very well may gain access to the web site of exactly the kind of company or companies you need. In any case I would recommend that, for your potential sub-contractor manufacturer, you locate manufacturers that make products similar in content and complexity to yours and that preferably are close enough for you to visit (repeatedly if necessary). Concentrate on ones that advertise that they do custom or job runs since they will be the most likely to be willing to listen and give you a price.

Now you probably will hit a slight snag. They may want engineering drawings (maybe even in electronically readable form) and materials specifications that you don't have a clue on yet. The closer the manufacturer's current products are to your new product, the less likely this will prove to be a sticking point. A model of your product and an example similar product may suffice. If you have a fairly complex invention and don't have understandable sketches, you may not be able to get any help from some firms. You also won't get much help if your invention, or your claims for it, appear to violate the known laws of physics. Again, don't be getting quotes at this point unless it can't be avoided. Also be aware that its is much easier to get plastic injection mold manufacturing quotes than metalworking quotes because differences in metal tolerances and finishes that you don't even know about yet can dramatically affect manufacturing cost.

Look Professional

Now is also a good time to look professional. Use letterhead paper, preferably with your company name on it. If you don't have a company (i.e., incorporated), use a DBA (Doing Business As) name. If you use anything other than [your real name] & Associates at your home address, your state and/or county and/or city will probably require that you file a DBA form. Call the county or city clerk's office and ask.

The fee is typically under $25—well under the $250 plus it might take you to become incorporated. If you do become incorporated you will also have to be mindful of the state and federal and sometimes city monthly/quarterly/annual paperwork requirements. (WARNING: A few days to weeks after you file DBA or incorporation papers you will start receiving solicitations in the mail claiming great deals on some of the things you will "need" as a new organization. TAKE THEM ALL WITH A GRAIN OF SALT. If you want to, file them, but when you are ready for such services, etc. do some good research and you will probably find much better deals.)

 You represent yourself as an "Inc., Ltd., Corp., or Co." without the legal entity paperwork filed. You use a DBA without a required city/county/state filing.

You do NOT have to get fancy letterhead printed at a print shop. You can get by with letterhead you design in your word processor on your computer and print on a laser or quality ink jet printer. From my own experience, I estimate that letterhead and typed requests will more than double the response rate you get on your request for bids.

I also find that the name and area of business you are in makes a difference. I tested Marketing versus Computer consulting once and got about twice the response for Marketing. I haven't tried White Industries, or

Worldwide Products, or whatever yet, but if I were interested in doing so I would very throughly check to be sure I was NOT using someone else's name. (See the information on Trademark searching for some tips on how to do this.)

Prepare your product concept fax package in advance. It should include a basic text description and drawings or sketches of the invention. Don't forget that drawings must include dimensions and (at least crude, i.e., "plastic") materials information. It is not always necessary to say what use the product is to be put to if it is not obvious, but it usually won't hurt either. Include your Non-Disclosure Agreement and (preferably with a rubber stamp) mark everything "CONFIDENTIAL."

Should you fax (or mail) your NDA first for signature and return to get a "legal" signature? My bias is NOT to be that paranoid. Very, very rarely would someone be asinine enough to question an exchange of signatures by return fax even after the information is sent. You should normally get a yes by phone to their general willingness to sign a reasonable NDA before sending your package but the mere fact that you included the NDA shows your intent to have the information maintained confidentially. Keep good records and build your body of evidence one document and note at a time. It is common for me to get "no bid" responses from firms that don't return a signed NDA. One day someone may try to burn me—and thus lose access to any future ideas I might have that could make them wealthy.

At least your cover letter should be on letterhead and it should spell out what you want (e.g., a quote for one or two specific run quantities, a quote

---

 You don't do any checking to see if the name you picked is free of conflicts.

---

on one-time tooling charges, help determining appropriate materials, a quote on product engineering, a quote based on specific product materials). If you want to look reasonably professional, be sure the fax is directly addressed to the person you talk to or the person they designate and that the person's name is computer printed or typed rather than hand written. It won't hurt to have a typist's initials at the bottom and a "CC: Jones Product File" (make up a better name than that please) either, but don't get carried away and create a pile of ~~lies~~ fibs you can be caught in either.

Lastly, as far as professionalism, do enough homework so that you at least come very close to getting the terminology of the industry that you are seeking estimates in right. If you request bids on a "spring" when what you really want is a "wire form," you may not only be looking unprofessional, you may be requesting bids from the wrong manufacturers. That will cut your bid responses down considerably and probably will result in higher bids than would otherwise have been the case. Is it a plastic extrusion mold or a plastic extrusion die you want? Do you want a circuit board or a backplane or a card?

The Fax (Mostly)

By letter, e-mail, fax, or phone (usually followed by fax), contact the prospective manufacturers, explain that you have looked at their ad or web site and/or products (if appropriate and be sure to name them), that you have a similar type product, and that you are in the early stages of development and need to get a preliminary estimate of WHAT THEY WILL CHARGE YOU to produce the item. If you want to, make them sign a non-disclosure form and let them know you have started the patent process (which is true if you have, for example, filed a Disclosure Document with the Patent Office).

The vast majority of the time you should be able to get quotes for free, however, if you have a complex(?) invention, and particularly if you have

---

 You make manufacturers deal by mail when they normally use faxes.

---

no mechanical drawings and/or electrical schematics, they may want to charge you $1,500 (or whatever) for an estimate because they have to do some rough design work. <u>That should not be considered unreasonable</u>. When YOU have real money and/or effort on the table you are likely to get a lot more help.

Quotes & Bids

Remember, you want to be nice to these people; during Product Development you may be able to get a lot of free advice from them. If you look like a black hole just sucking them dry (like the other 3 inventors that approached them this week), you many not get what you want or you may have to work extra hard to get it. Some manufacturers may give you free estimates even if you don't have a clear design (particularly for simple inventions) in the hope you will work with them, and perhaps pay them for their help during development, and subcontract at least your early runs to them.

If their estimate is way off from your estimate based on the preceding division-by-ten-then-apply-a-multiplier, get another estimate or two—you may have settled on someone specializing in prototyping only or really short runs or picked an unjustifiable multiplier yourself.

DO NOT request a detailed, fixed-in-concrete and good for 2 years, quote. (Then they will know for sure you are a black-hole—and a wannabe.) You may ask them for estimates on 2 different run sizes though, say 5,000 and 25,000 or whatever might be appropriate for your product and initial realistic sales expectations. You'll have to use some judgement based on your invention and the probable number of custom versus stock parts, but I would think expecting minimum runs of under $10,000 for anything moderately complex (like half of a toaster) would be unrealistic unless you were dealing with a specialty producer geared only for very short runs (and charging per unit prices commensurate with that). For example, I have a price of $1,000 for 10,000 of a single bent wire item and another of $6,000 for 2000 of an item with 5 plastic parts (excluding assembly). Be thinking in terms of first PRODUCTION runs and excluding tooling charges, not PROTOTYPE runs which you will probably also have to pay for but are not anywhere near ready for yet.

I saw a classic example of the results of NON-PROFESSIONAL requests for pricing on the Internet recently. The inventor was whining to the whole forum that the price of the ⅓ to ½ horsepower electric motor that he needed to complete his power saw invention was way too expensive at $35. Could anyone suggest a different source? Reading between the lines it was clear that the inventor had called the manufacturer and asked for the price of the motor. And that is exactly what he got: order one and you can have it for $35.

I bet if he took the trouble to prepare a request for quotes (RFQ) for 1,000 of that same electric motor it would likely cost about $12 each. Now we're on to something. Make that a guaranteed 100,000 motors and the price will likely be under $6. Now we're getting down to where component costs will be low enough to let us sell at a profit IF OUR PRODUCT WILL SELL AT A CONSUMER'S "REASONABLE VALUE" PRICE. Another example, "fluorescent" type super bright white light LEDs: each @ $12, 10,000 @ $6, and 100,000 @ $1. If you want volume prices you have to be prepared to buy in volume.

I'm an OEM?

Yes, you are an OEM! (You might also be a DOM or an SYT but those are not relevant to the context of this book.) You are an Original Equipment Manufacturer even though you will be subcontracting all or parts of the manufacturing. Dell, the computer company, for example is an OEM even though they mostly just custom assemble computer parts made by others. You should plan on building based on their very successful model.

You should also be aware that OEMs get the best prices from other manufacturers WHEN THEY BUY IN OEM QUANTITIES. When you contact manufacturers of off-the-shelf parts or materials you should always state that you are a manufacturer and ask what their minimum dollar or quantity order is to qualify as an OEM. Remember to be professional with letterhead. If you cannot afford the minimum you will probably have to go to a distributer where you'll get a lot less for a lot more.

For example, OEM quantities of wire are typically 100 lb spools and you might easily get a lot of 10 spools for under $300 per spool plus shipping. That is a fantastic deal compared to a distributor price on the same wire in a 7 lb coil at $15.19 per lb plus $9.60 for shipping. For my prototypes, by the way, I didn't order either quantity, I "let" my wire form maker provide the closest available wire from their own stock. The first production round will probably be via a distributor but after that the OEM quantity from the wire maker will probably be the way to go.

A rail car load of plastic is probably beyond your OEM prototype needs too, but for production of the right product it won't be. I've picked what you might consider "giant" quantity examples here but you often don't need tremendous volumes for OEM pricing. You also need to pay close attention to what you are getting quotes for. Development and initial production quantity quotes will be highly relevant to your start-up, but real OEM quantity quotes should appear in your business plan costs as early as is realistically practical.

Can't (Or Won't) Get Manufacturer Quotes?

What if no manufacturer will give you a quote or you think you have valid reasons for not contacting a manufacturer yet you feel you need a quote? You can buy one from a service that specializes in estimating production costs. Or you can go to your nearest engineering college library and ask the reference librarian there what pricing handbooks or software they have. If you have some knowledge of the field, you may be able to use them to come up with some reasonable numbers on your own.

A couple of places to start are *Journal of Manufacturing Systems*, Vol. 7 (1989), No. 3 pages 183-191, "Early Cost Estimating in Product Design" by Dewhurst and Boothroyd (this is high level) or the *Handbook of Electronics Industry Cost Estimating Data* published by John Wiley & Sons. (I've only seen a 1985 version but I don't believe there is any newer one.) The *Handbook of Electronics...* has a decent section on sheet metal working operations but its section on plastics only notes the cost of tooling ($10,000+, i.e., about the same as today). While the electronics parts prices are out of date they may actually be higher than today's prices. Also, newer technologies such as surface-mount chips simply aren't there.

There are similar handbooks for metal fabrication and plastics but you'll have to work to find them. When you do find one you will often be confronted with the need to know machine power ratings, feed speeds, distances between tools and other such stuff that may make you realize estimating is not your game. Sometimes you can glean information from a college text book such as *Manufacturing Processes and Materials for Engineers*, Lawrence Doyle, 1969, Prentice Hall.

Some manufacturers of the equipment that is used by an industry also have "books" that suggest appropriate estimating methods for use with their equipment but you'll have to ask around to find something. Steer clear of the books that emphasize process design and plant construction. Big firms, and even some small firms have there own estimating databases but the odds of your getting any access to them are slim and none. Software is also available for estimating.

A couple of places you might try first are: <u>engineering.software-directory.</u> <u>com/data2/cdprod1/doc/software.category/Engineering.Construction..</u> <u>Estimating.and.Costing.html</u> (note the double dots and capitalization), a site by TurboGuide, which lists over 60 estimating software packages (but not just parts production estimating). The best software they probably list is Costimator by Manufacturing Technologies, Inc. (<u>www.costimator.com</u>) but you'll also find that their software pricing structure ($5,000 to $50,000) is geared to industrial buyers. The other place is <u>bizweb.com/keylists/</u> <u>manufacturing.software.html</u>, by BizWeb, which has an even longer list of software to sort through.

Real quotes are always best but if you must be ultra secretive the "find similar products and divide their prices by 10 then apply a multiplier" method will probably suffice and be far easier than trying to research or understand industry costs. You can also use Ask Jeeves or other search engines to search the Internet for fee based estimators that you must pay.

Unless the fee based estimator is affiliated with a manufacturer you must understand their numbers are not binding quotes. If you get an estimate from one of these outfits and it later turns out to be way off (and your invention has NOT changed from what you gave them to estimate) recent case law suggests you may be able to successfully take the estimator to court and collect damages IF you made decisions based on their estimates. It's been said there are 2 ways to lose a law suit, one is by starting it and the other is by giving someone else an opportunity to start it. Remember, you shouldn't even need to be getting estimates if you reasonably believe your division-by-ten-then-apply-a-multiplier effort gave you a reasonable answer.

Evaluate—Again?

Now we have some information to analyze, including: competing products, BFH solutions, snicker test results, a marketer's evaluation report (hopefully with a suggested selling price and long term volume estimate), one or more production cost numbers, and a buyer comparison expectation price. If you got manufacturer quotes then I would suggest you start to sort it out by taking the highest <u>production</u> cost estimate from a manufacturer that you believe to be reliable and multiplying that by 5. If that is considerably higher than your marketer's suggested selling price and/or the buyer comparison price you determined, it may be time to stop and start on your next idea. Such an "overpriced" product can only likely work where you will be able to get patent protection and where you will be catering to a market whose problem costs, or lack of competing solutions, make your product attractive at a higher sales price.

Suppose the price check comes up near or below the proposed prices? Next look at competing products. Is one or more of them likely to be just as effective as yours but at a substantially lower cost? (If that is the case then STOP, you will NOT be rewarded for YOUR cleverness—the buyers vote with their dollars in the way that best maximizes their VALUE from the solution.) Is a BFH solution workable in most real-life situations? Will consumers probably perceive your product as being more beneficial, in their terms, to them than competitor's products? Tough questions to be sure. What do the results of the snicker test, weighting the expert responses most heavily, have to say about these questions?

Quick & Dirty Business Plan

Draft a 3-5 Year Business Plan Spreadsheet

If everything still looks like a GO at this point you should do one more "little" analysis exercise. Do some careful thinking and do a quick and dirty projection of numbers for a (possible future) business plan. Using projected volumes, conservative market penetration rates, the cost of any loans you may need (or the percent of profit you must give up to your financiers), marketing costs you project (based on your average competitors marketing costs times 5, or 15% if you have no evidence for a different number), shipping/warehousing costs, the subcontractor-manufacturer's price to you, packaging costs, liability and insurance costs, and all the other costs, see what your projected profits will be each year for 3 to 5 years. (See Chapter 8 Marketing 101 for help on finding numbers.)

NOTE: Some people are remarkably good at BOTEC (Back of the Envelope Calculations) or even mental profitability evaluations. They generally just seem to have a knack for it and a very realistic approach to numbers. I applaud those kind of people. However, most inventors I've met are terrible at that. Estimating 100,000,000 sales the first year in the U.S. or 2 billion worldwide magazine subscriptions come to mind and are obviously NOT realistic.

On the other hand, I expect that once you have carefully worked your way through the numbers a time or two and get a good feel for what the census and other numbers are and mean, you too will be able to quickly filter out ideas with minimal probabilities of success.

If you cannot find published numbers for your target market be very very cautious. The odds are you are trying to identify prospective buyers that are so diffused throughout the population that the economics of marketing to them will limit your chance of success. If you cannot find the numbers, ask yourself how you will find the people—and what it will cost to do so. Certainly, if you set up a web site, some will eventually find you—but will it be enough to pay for the product plus your marketing and overhead? You must build your marketing costs into your product costs. **In business, the marketing costs do not come out of profits.**

Be aware that marketing costs will be dramatically higher for a startup than for an ongoing operation and that true marketing costs are often buried in the cost of goods sold figures for firms that use manufacturer's reps and/or seller-distributors. <u>If spending a day or two or more doing the numbers for a quick & dirty business plan is too tedious for you, the odds are against YOU having the ability to set up a day-to-day company that will have long term profitability.</u> You still need to do it, but you may want to put your numbers in terms of a 5% of manufacturer's price license and show how the licensee will profit from the rest of the numbers. Or you may want to do the numbers in terms of a <u>paid manager</u> (salary and/or equity) for setup and running of the company.

I strongly suggest you do your projected sales and profits exercise with a spreadsheet program on a computer. Otherwise, the math will get so tiring

---

 You don't do any kind of profitability or payback forecasting.

---

you'll be tempted to make oversimplifying assumptions (usually toward inflating profits). If you don't identify at least 10 variable and fixed expenses beyond the cost of the product itself, you are probably doing something wrong. Get help from the Small Business Administration's SCORE (Service Core of Retired Executives) program (www.sba.gov) or a friend or a neighborhood business college student or even a business plan book or software package. DO NOT expect SCORE volunteers or any of these other people to do the work for you unless you are paying them.

WARNING: Don't guess at numbers unless you have to. Do some research and be able to justify your numbers. For example, the 15% suggested above is a broad average for startup marketing of retail products. Windows 95 had startup marketing costs of $200 million—nice if you've got it! If your product leans toward the buy-it-when-needed commodity type (e.g., nuts and bolts), you may be able to get away with lower marketing costs, especially AFTER distribution acceptance, but if your product leans toward spur-of-the-moment (e.g., exercise device), you may need considerably more to get a sufficient consumer mind share for your product to be chosen over competitors' products. If you're going for a 1-2 year fad product pray for favorable publicity.

Also keep very good track of where you got your numbers and document them very carefully. The documentation will be essential to preparing a real business plan when you need it to convince financiers or other business people of what your product can financially do for them.

Now, after seeing realistic profit projections for at least 3 years out, make a decision based on how willing you are to keep your "day" job and work your new inventor/manufacturer "job" for as long as it takes.

Success Stands a Chance

Congratulations, by now you have winnowed your first hundred ideas down to one that you feel you are willing to risk some serious time and money on just to get it to the marketplace in hopes that it will be the one out of the estimated twenty simultaneously newly introduced products that actually succeeds. The first hard part is over and you can proceed to the fun part of inventing with renewed vigor and confidence. You might have a winner.

12. Your average comparable manufacturing complexity price (before dividing by 10) is significantly higher than competing products with results similar to those of your invention, OR...

 **IF**

13. Multiplying your highest reasonable manufacturer quoted price times 5 produces a selling price significantly higher than a buyer is likely to expect or accept...

 AND your invention, based on the "snicker test" and expert opinion, has no strong benefits that can overcome high pricing.

 AND you won't be able to get patent protection where the costs of even an expensive solution make your invention attractive.

14. Your (quick & dirty) business plan shows payback will take longer, cost more, or be less substantial than you are willing to accept.

Explanations and scoring suggestions:

12. <u>Your average comparable manufacturing complexity price (before dividing by 10) is significantly higher than competing products with results similar to those of your invention.</u> You should have been able to look at products whose materials and manufacturing processes are similar to yours. If for some reason that wasn't possible, you can score this issue a 0% and provide a percent estimate for issue (13).

 0%—Your invention's manufacturing complexity, and therefore probable production cost, will <u>definitely</u> let you price your invention <u>well under</u> the competition.

 5%—Your invention's manufacturing complexity, and therefore probable production cost, will <u>probably</u> let you price your invention <u>well under</u> the competition.

 25%—Your invention's probable production cost will let you price just under the competition and other factors don't give you a clear edge over the competition.

 25%—Your price will have to be high but you can get patent protection and your solution is still very cost effective.

50%—Your product's probable price will have to be a bit over that of the competition but other factors may entice buyers.

 75%—Your product's probable price will be above competitors but you <u>may</u> have some possible buyer advantages.

 75%—Your price will have to be high but your solution is very cost effective, however, you cannot get patent protection.

 100%—Your product's probable sales price will be well above competitors and you provide no special benefits to the consumer.

13. <u>Multiplying your highest reasonable manufacturer quoted price times 5 produces a selling price significantly higher than a buyer is likely to expect or accept.</u> If you didn't get manufacturer estimates of production costs because you did an off the shelf product production cost analysis, you can give yourself a 0% (☺) for this issue. If you did get a manufacturer to provide estimates then use the scoring suggestions for the previous issue to score this issue.

14. <u>Your (quick & dirty) business plan shows payback will take longer, cost more, or be less substantial than you are willing to accept.</u> The more carefully you estimate costs, sales and sales growth, and market size, the better your chances of becoming a viable long term business. If you must or prefer only to license then figure your license percent conservatively as 1 + STEP # completed prior to licensing (e.g., 6 if you complete STEP 5-Sell a Few).

 0%—The probable profits came out higher than your wildest dreams AND you are very certain you were conservative in your estimates of market penetration rate and market size.

 5%—The probable profits come out high enough you could easily hire professional management and move on to your next invention.

 25%—The expected time till payback exceeds 2 years and you DO have the money or financing to hold out that long.

50%—The profit won't let you quit your day job.

 70%—You project a good profit but your market size or market penetration rate may be optimistic.

 70%—To get financing for a 3+ year payback you'll have to give up control AND 70% of the amount to be overcome is one-time startup costs.

 75%—You didn't do a careful costs and receipts analysis.

 100%—You project losses or minimal profits.

 100%—You believe everybody will buy one, maybe even every year.

CHAPTER 5

Steps for Product Development

Now, based on what you know from your Focus Groups, Mall Interventions, etc., that were done while working the idea through some basic "Will it sell?" market considerations, it's time to get serious about development. You're no longer just guessing that your invention has a chance so financial commitments make good business sense.

Have Some Fun

Mr. Merrick says two things on page 60 of his book which I will repeat verbatim:

1. "Inventing is the easy part of turning your ideas into dollars. But, after you think up a brand new product, there's a tough war you must fight for market acceptance and product distribution."

2. "The Stand-Alone Inventor is an individual who not only does the inventing—the quick and fun part—but also starts a company and learns how to get the manufacturing and marketing done—the long-hours, not-so-much-fun part—but usually the part that makes the money."

I think he is pretty clear and I agree with him. Since the intent of this document is to focus on the marketing aspects of the initial stages of inventing, this section will be a bit shorter than the previous one. Take your potentially successful idea now and proceed with STEP 4 which is part of the fun part of inventing.

Licensing Alternatives

If you are not interested in being the manufacturer/seller of your product and don't want the fun of development or expense of patenting here would be a good point to concentrate on licensing your invention. At this point you do have cost and revenue numbers so you should be able to intelligently and realistically negotiate a fair licensing agreement. You might also want to consider filing a Provisional Application for Patent (as described later) before starting licensing negotiations but it is not required. You could, if tinkering is your joy, continue through STEP 4 and get a provable working prototype finished before starting licensing negotiations also. And if you really want to be in the best negotiating position with prospective licensees, you will continue through STEP 5 and prove the product will actually sell.

STEP 4—Fully Design and Refine Your Product.

Patent considerations for this step: Get non-disclosure agreements signed by anyone you work with on development. If it is a manufacturer, or even an individual, whose expertise contributes to executing your ideas, I suggest a statement like "Any improvements, whether patentable or not by [manufacturer or assistant name] shall be owned by [your name]." If it's friends (but not inventor's notebook witnesses) or inventors club buddies you probably should work out a sharing agreement in advance and PUT IT IN WRITING. Include what happens on death: does the spouse get the share (all or part?) or is it split among survivors?

Don't go to your patent attorney until after your invention is pretty close to final as shown by your working prototype. At this point you only want to get your attorney familiar with your invention and get you familiar with the possible costs and patenting processes and issues (more info on this appears later in the text for this step). In my opinion, it is still not time to file a patent application.

If you filed a Disclosure Document you hopefully will still be well ahead of its 2 year destruction date. If you launched the patent application earlier anyway, now is probably where things get embarrassing. You discover your invention doesn't work quite like you thought it would. You can't add claims to your current patent application so you have to pay your attorney to file a new application or a "Continuation-in-part" for more costs and fees. You also realize you may now have to pay to keep 2 patents, IF you get them, in force.

Simplify

I suggest you work with pencil and paper first and do several iterations of quick and dirty designs. Do them preferably in your inventor's notebook but be aware that erasures, whiting out, and other kinds of "corrections" can make the contents of your notebook suspect so don't make corrections—start the drawing over. You can also do drawings on other sheets of paper and attach all or representative ones to your notebook pages or you can work with pencil and paper then draw current results in the notebook. Keep everything and remember: all you are doing is building a body of evidence, the more you have the more effective it will be. For more information on keeping records see the Nolo Press books or get a copy of *Idea Journal* (Ideastream, Inc., 11410 NE 124th St, Suite 216, Kirkland, WA 98035) which is bound inventor's notebook with instructions for proper use.

For most people paper and pencil sketching will probably be faster than trying to do engineering drawings, even on a computer. If it is somewhat complex in the user interface you may want to do some drawing in a computer drawing package because that will make moving around and re-labeling user interface components easier. Pick whatever works for you but assume you will radically alter your basic design as you progress.

When you get your quick and dirty design simplified to where you believe it can be simplified no more, stop and re-look at the problem that you are solving. (Maybe you should have done this as part of STEP 0, but I'll leave it here because I suspect that your problem statement for STEP 0 would probably have best matched other people's perception of the problem and restating it might have just been confusing.) Try stating the problem in several ways to see if it changes your perception of what "the" (your?) solution might be. This can be difficult but I will give you two examples that may get you started.

Try Restating the Problem

The first is a problem that Bob Merrick solved. Bob's (and I'm putting words in his mouth) problem statement was "How do I punch Rolodex type notches in business cards so I can put them in the Rolodex instead of copying their information to pre-notched Rolodex cards." His solution, available directly from him for $7.95 plus $3.00 shipping and handling, is the Personal Punch business card punch as described in his book. His solution also was not complete in that users were cautioned to hand copy material from where the notches were to be punched. In fact Bob created a national organization whose purpose was to standardize business card layout to avoid the notch punching area.

What if the problem were restated as "What can I make that will hold a business card in a Rolodex without my copying its information?" The solution to that problem statement exists also: it is a pre-punched semi-rigid clear plastic flip open "Name Tag" type holder without the wire pin for attaching it to your shirt. They work but some business cards must be trimmed or partial notches punched (with Bob's punch?!) and care must still be taken to avoid losing information.

One more try. What if the problem were "How do I get notches onto a business card so it can be put into a Rolodex without copying information?" John Merrick, Bob's son, solved this one with Valuable Contacts tabs. You can buy a box of 100 semi-rigid clear plastic self-adhesive "tab" strips each with 2 properly spaced Rolodex notches for $7.95 (plus $3 S&H) from Bob.

Bob's punch is buy once, use forever; the "Name Tags" are buy a bunch, re-use forever, John's tabs are buy many forever. I must confess, I haven't tried John's solution yet, but I like it. All solutions available at well-stocked office supply stores, prices may vary.

On the other hand, I'm so confused (???...) I might just staple or glue or tape business cards to pre-punched Rolodex cards. I haven't tried those "BFH" type solutions yet either. I checked with Bob and he says his and his son's solutions are selling neck and neck now and he expects them both to continue for a good while.

The second problem is for the toilet seat lifter/lowerer. (WARNING: If you are politically sensitive please skip the next 3 paragraphs. If you do not skip these paragraphs NOW I refuse to be held accountable for any resulting emotional or health problems.)

Various versions of the problem statement: Men don't put the toilet seat down after standing to pee (men are insensitive clods version). Women don't see that the toilet seat is up (women are not competent to look out for their own safety version). How do I ensure that the toilet seat is only up when a man is peeing standing up (up for standing peeing otherwise down version)? How do I ensure that the toilet seat is down when anyone wants to sit on it (down for sitters version)?

Solutions: Increased sensitivity training at "guy" school (problems: missed day at "guy" school, inadvertent error). Light attached to seat that glows red when the seat is up and green when the seat is down (problems: color blindness, failure to associate red with the requirement to lower the seat). Seat lifter/lowerer with lift activator at toe position of standing man and seat lowered after toe removal (problems: accidental heel activation by sitter causing unwanted pats on the bottom or bouncing during business, toe removed when dodging a sidewinder). Press bar activator (to lower or rotate or slide or whatever the seat into position) high enough to clear male appendage but low enough to be contacted by back of sitter (problems: bar too low for extra tall guys, bar missed by doubled over sitters and small children, possible violent seat removal at worst possible time—female cramp episode).

My solution: I always leave the seat down after I'm done—why risk the (irrational?) ire of a woman when it is so easy to avoid? Problems: not patentable, only 99.99% effective for me, works only in my realm in my presence.

Evaluate Yet Again

Now is your solution still the best (simplest, least potential development pitfalls, etc.) that you can come up with? If so proceed, if not, consider the market share that you might garner for your current solution versus reworking while you are still in paper development mode. Committing yourself (and your ego) to a solution too early in the process is a guaranteed way to waste your time and money. If Bob Merrick's Personal Punch had been as complex and expensive to build as those of either of his eventual competitors, even assuming he had gotten patent "protection" for the complex version, his market share would have evaporated when someone came out with one as simple as his finally was. I repeat, a patent is not for an IDEA, it is for a PARTICULAR EMBODIMENT (usually) of a problem solution.

I should also note that I'm assuming that you spend a week or less at your restating and rethinking. This is particularly important when you are looking to enter a newly growing market niche such as bicycle carriers for SUVs. You can't spend 3-6 months toying with alternatives, you have to get something to market before you get drowned by the competition and you can refine it later. Still, a couple of days to get it close to "right" the first time is worth tons of fixing farther down the road.

If you achieve a simpler solution than the one your potential subcontractor-manufacturer(s) estimated for you—rejoice—you are probably gonna make a bigger profit than you originally thought. (Of course you do run some risk of having already rejected working on a problem for which you might have eventually hit upon a better solution. Don't worry, it still may pop into your head at the least expected moment. Your subconscious probably has not forgotten the problem.)

If you diligently did your "seek alternative solutions" and patent searches and asked qualified people before, and they didn't route you to an existing product like your new revised one, the odds are probably against your finding one on the market already. If you believe it would be prudent to double-check, then do so while your costs are still low.

Why do you want to go through the above exercise? Why not just accept your invention as is and go? Because you want to avoid getting killed in the marketplace by a big company (or another inventor for that matter) who sees

You don't go to the trouble of simplifying your design and thereby leave the door wide open to competitors.

your solution, surmises that it is selling profitably in its market niche, and creates and markets a "better" alternative for the same niche market in 6 months or so. A problem can go unsolved for years simply because nobody recognized a viable market, but once that market is discovered there will be many people who want to see if they can cash in on it. (Witness the "Titanic" feeding frenzy, that's not an invention example of course.)

If you are willing to pay $250 per hour for my time I will be happy to come up with alternative problem statements for you and even help brainstorm for solutions. (I only want the hourly money. You can keep most of the credit even if it is my solution you develop, patent with me as a co-

You are impatient and leave all the detail problems to your paid help—and co-inventors.

inventor, or the only inventor, and sell.) You can do the same with a friend but be prepared to reward them lavishly when you strike it rich. Be prepared to give up half your profits (plus some to legal fees) if it is their solution you ultimately develop and market. Also remember that simple means simple.

If your design keeps hitting snags that need more and more complex "little" solutions to solve them, it is time to rethink the basic solution. Also keep in mind use of the invention and how safe, simple, and easy it will be. If your invention is simple—but requires a special additional gizmo or other product for efficient/effective use—do you think your solution is really going to be perceived as a solution by the buyer? Maybe, but then again, probably not—or at least not in a big hurry. Electric light bulbs needed electricity, televisions needed broadcasters, so even though they were overnight sensations, success took many years. Solutions that spawn problems are unlikely to be long term successes. Evaluate carefully.

A Model

Now you are getting close. You may have had a model or a mockup to show in earlier steps but you probably shouldn't have killed a lot of time or money on it unless that was truly essential for understanding by prospective purchasers. Now you are going to need something that adequately conveys

---

 Your "simple" product requires some additional things for typical use.

---

the information needed by an engineer or prototype builder (if you are getting help). Your choice is whether you build a model (my recommendation) or detailed sketches for a machinist (or whatever) to build a prototype from, or do detailed engineering drawings and seek assistance from one of your potential subcontractor-manufacturers.

A model or a detailed engineering drawing (especially on a computer/software system that can test assembly) or even creating a prototype (assuming parts are made nearly identically to final manufactured part designs) will often turn up little details that need to be dealt with. For example, it is physically impossible to assemble or cannot be removed from a mold in its current design incarnation. The bane of sophomore engineering students is to lose half credit for a technically perfect drawing of a device that can't be built or that would be prohibitively expensive to build (e.g., assembly with a split metal ring that must be welded and machined to smooth circular perfection for successful operation). It is best if you turn up those kinds of problems yourself first before getting someone else involved.

Cardboard or paper models with tape and glue need to be reasonable but not necessarily precisely like the final product (e.g., simulate gears with toothless disks having a bead of rubber cement instead of gear teeth). Spending weeks creating a near perfect model is probably not a good use of time. What you want is a model that accurately conveys what is to be built. If you can build the prototype yourself that is ideal—but if you have to buy lots of tools and/or equipment and learn how to use them proficiently—forget it, pay someone.

Liability

If you know or suspect that there will be some liability issues involved with your invention, now is the time to start dealing with them. Some really dangerous products, such as toys, (yes you heard that right) have bodies of law with safety specifications that are mandatory for you to meet or exceed. It certainly won't hurt you to check out the Consumer Products Safety Commission at www.cpsc.gov.

Go in and rummage around and do searches of the Code of Federal Regulations for your product or product type. You'll probably notice the CPSC site occasionally bounces you out to the www.gpo.gov and other sites.

 You don't think "they" can do anything to you if you ignore safety issues or liability.

From the CPSC you can request a free copy of *The Handbook & Standard for Manufacturing Safer Consumer Productions* (June 1975). It describes at a high level the issues you need to think about from the day you start working on your invention. **If you don't have time to look at the CPSC site now you can always come back to it later to find out how to deal with a mandatory product recall or a violation notice.**

While you are thinking safety, don't just look at the laws. Many safety standards that you MUST meet in order not to be easily found liable are "voluntary." Another government site good for gathering this type of information is the National Institute of Science and Technology (NIST, formerly called the National Bureau of Standards). Their web site is www.nist.gov. You can collect a wealth of information either online or by ordering documents.

You will also discover that there are many independent testing laboratories that will "certify" a wide variety of products. For example, I did a search and found 3 labs that certify toys. Yes, this stuff can be a lot to wade through. **A shortcut is to buy a few products related to yours and closely examine them, their packaging, instructions, manuals, etc., and see what kinds of certifications or standards they purport to meet then to specifically research those.** If you plan a global product the International Standards Organization (ISO) site at www.iso.ch is probably also a must visit. Their online catalog, for example, identifies a dozen publications relating to metal color and fineness standards relevant to jewelry products.

One typical "instant" reaction to the problem of liability is to incorporate or establish some other form of limited liability organization. It is a good idea but only to a point so I don't see its necessity in the early stages of your firm. I do highly recommend you get additional advice on this though.

The reason I'm not so keen on immediately incorporating or whatever is because both a financial and a time burden are imposed on you by such action. You have to keep the entity's records separate and apart from your own. That usually requires paid people or a considerable learning curve on your part.

No matter what you do, however, if your assets (or earning power) are bigger than your company's, and it appears that the event or decision for which the corporation "is" liable arose from negligence or a decision on your part, the plaintiff's attorney will quickly "pierce the corporate veil" and zap you directly. This is pretty much 100% guaranteed if your "Inc." doesn't carry a reasonable level of liability insurance for you.

Finally, a Prototype

I recommend you do as much as you can to complete the design and refine your invention yourself before getting outside (usually paid) help. You may need to research what standardized parts are available for some things (e.g., nuts, bolts, push buttons, gears, shafts). Using a readily available 2-cent part is usually preferable to using a 10-cent (excluding one-time charges like tooling) custom part. Your best source for this kind of information might be the machinist or electronics shop that will do your prototype, but it is more likely to be the engineers at your subcontractor-manufacturer's site.

If you got quotes from manufacturers in the previous step hopefully at least one does prototyping, better yet rapid prototyping. If none of your quotes are from manufacturers that do prototypes you should find one that does or find someone that specializes in building prototypes. Now you can ask for a nailed-down quote for just the prototyping if you didn't get one already. If you were able to skip bothering manufacturers for quotes in the previous step you should go back there now and review how to professionally request a quote (pg. 81). Once you have one or more quotes, gulp!, you'll have to commit to spending some real money.

Yep, at some point you need to get your first prototype built. Expect to pay for it. I spent over $300 for a machinist to create the working parts of my first (unpatented) invention. At the time, of course, I thought it was a

---

Fixes to the design during prototyping turn up requirements for special materials or expensive processes.

---

fortune but in hind sight I realize it was a trivial amount. Try the prototype and see if it works. <u>You will probably be embarrassed to discover that it really doesn't work to your expectations</u>. Time to have some fun and refine and retry till you get a prototype that works really well.

CAUTION: watch your refinements to be sure that final manufacturing costs won't knock your product back into the unprofitable zone. Needing special materials or time-consuming machining or any of a host of things may mean it is time to stop and rethink—maybe even stop entirely. If you want to be a <u>financially successful</u> inventor, don't forget that the market does not reward people who successfully solve a problem—**the market rewards people who successfully solve a problem at a cost (to the buyer) lower than the value of the solution (to the buyer).**

This part of the process should make it clear why Mr. Merrick and I and many others recommend that you work on your best SIMPLE ideas first and leave the more complex ones for later (if ever). If you don't know how to do something or lack materials expertise—get (paid) help—you'll waste much time, if not much money, reinventing something that you can't patent or otherwise protect.

Rapid Prototyping

A word of caution here. Where possible I highly recommend some form of rapid prototyping that gets you as close as possible to the final manufactured product. Demand that you own the engineering CAD file (AutoCAD compatible) for any engineering drawings you have done.

Manufacturers will probably need 3D "solid" design (.STL) files rather than older 2D style "front, top, side" view style files. That lets them drive their machines directly from the computer but it also makes it possible for you to get nice printouts of your invention as viewed from different angles. Especially if you are paying the full cost of the CAD work, make getting the file a simple requirement or they don't get the work. Also don't hesitate to let them know that as long as they continue to give you a fair deal you will continue to work with them.

You will undoubtedly have gotten, or will get, quotes from firms that have their own way of doing things. If that meets your requirements, you can work with them. IF IT DOESN'T, go somewhere else. For example, I have an invention I want done in plastic, but while the first prototype maker I went to does do work in plastic, they could not do it as thin as necessary for the real product. They suggested I have them do the prototype in metal which they could work to the correct thinness.

To accommodate metal construction required significant differences in the design. Design and first prototype would be $5,200 and subsequent prototypes would be $60 each. Of course I turned them down. My all-plastic, nearly-final prototype maker only quoted $2,600 for the first prototype and $400 per hundred thereafter. Other situations can arise where the opposite is true, i.e., it makes more sense to do a prototype that is not of the same type of material as the final product will be.

Problems with Manufacturers

Another problem you will encounter sooner or later is that your product or its parts easily fall within the scope of the technical specifications of the equipment that the manufacturer says they own. BUT, it turns out, that while you're within the range of two specific specifications separately, they both can't be met in the same part. (For example, a .005" diameter bend .05" from the end of a piece of wire.) So much for the value of the machine manufacturer's claims of 6 axes of motion or whatever. Sometimes you may find that other manufacturers of equipment do not have the same limitations or you may find that the only way to try to get what you want is to use whatever program for prototyping the equipment manufacturer might have available.

Since this has happened to me, I have concluded that the most effective approach might be to challenge all of the appropriate equipment manufacturers (find them in Thomas' Register) to make your part(s) and see just who will work with you. Remember, resort to this only if something that you are pretty certain should be manufacturable isn't on one or more machines. By asking around you may get referrals to a particular manufacturer that is known for handling particularly tough problems in a specific area. You will very likely have to pay more to this manufacturer to get as close to what you want as possible. **Try to figure out if what you want is necessary to your customer, or just bull-headedness on your part, before getting too stuck on it**.

Some manufacturers may want patent rights to any technology they have to create to meet your requirements while others may just figure it's an "obvious" extension of existing technology. If they want to patent the technology or process necessary to make your product, you should either run like hell or sign an agreement specifying exactly how their patent rights will (or will not) affect your product cost depending on how much you trust them. The last thing you want is a manufacturer who can rape your profits because their process or technology is essential to your wildly successful product. It would, of course, be better if you could invent the machine or added features or the process needed for making your product, but I suspect that, time and cost wise, is impractical for most inventors.

Yet another problem you'll now encounter is speed—or lack thereof. You are all charged up and enthusiastic and have been working full-speed-ahead on your invention for some time. But the people you pay to work for you, from draftsmen to engineers to prototype makers to manufacturers, will probably not share your sense of urgency. In fact, they often won't meet the dates before which they promise to have your work done. Multiple iterations of something that could be accomplished in weeks typically drag out over months.

This lack of speed is another very good reason for you to have worked your invention down to what you think is about its best possible form before getting paid help. The catch, of course, is when you need expertise in order to close in on that ideal version. Do your homework and be able and prepared to ask enough questions and provide enough information that your paid help won't have to proceed too slowly for fear of misunderstanding. Think of the time you must cool your heels and wait as an opportunity to consider marketing issues or even to plot out your next invention ideas.

One last catch with manufacturers. Ask to see some of their product output right up-front. Does their quality meet your standards? If none of it does you probably should go elsewhere. If one or two items do, select those and specifically tell the manufacturer that is the kind of quality you are looking for. **If you build your quality requirements in from the start you and your manufacturer will have a better experience and it will cost you less in the long run.**

The Cat's Meow

It is best if you do not keep thinking up refinements for your product, you were supposed to have that pretty well completed in the paper and pencil stage of design. <u>DO NOT keep going back to your help with more and more change suggestions or new questions while they are executing what you already asked for</u> or you'll be the one to blame for delayed delivery and higher costs, not them. Your objective is to get a prototype of a marketable product. Almost all products that sell go through further development and enjoy additional releases of enhanced or refined versions.

New automobiles come out every year. New software releases come out on an irregular basis. New editions of books are printed. Fashions change. Color tastes change, etc. Your product has to work but it doesn't have to be

---

 You delay introduction to add new improvements.
You leave problems important to the buyer unsolved.

---

the cat's meow yet either. Refinements will be much easier with profits in hand also. In other words, develop to a point of a viable product then

FREEZE the design and features list and concentrate on taking that version of the product to market.

Of course there is a gotcha. The marketed product has to reasonably meet the expectations of the buyers for both function and quality. Only in the computer software world are we completely trained to buy each "next" version when it comes out in order to escape the problems of the current version. You probably won't be so lucky with your product. If you (via your product) are NOT giving buyers what they expect in terms of value, they may knock your product to all their friends and thereby kill off any chance you might have had at long term success.

The "new, improved" version 2 may simply be rejected because of your reputation even if the new version fixes all the problems with version 1 and adds some fantastic new features. Like they say, "You never get a second chance to make a first impression." But you can't let fear of rejection paralyze you into inaction either. Sigh, why can't this be easier?

Cost Plus

What if all manufacturers you approach for prototypes see possible problems such that they are not sure they can accomplish what you want and therefore won't bid on doing the prototypes. You may be able to find one that will work on a cost plus basis. On a cost plus basis, the manufacturer will charge you direct labor, materials, perhaps machine time or other direct charges plus some extra amount (usually based on a percentage) to cover their overhead (and perhaps yield a small profit). The "plus" could also be an all-or-nothing amount paid if the prototype is made successfully.

You need to evaluate this option carefully and work as closely as you can with your manufacturer or prototype maker. It is best if you figure out in advance some kind of level at which you will say "when." Have the manufacturer keep you appraised of their estimates for production runs for whatever complexities are causing problems. When you're reasonably certain you've hit the "overpriced" zone you computed in the previous step, don't hesitate to pull the plug, pay your bills, and get serious about your next idea. Sure it hurts and yes you've lost (relatively little?) money, but that's business.

Design & Engineering

I have intentionally de-emphasized the design and engineering aspects of product development because they are NOT what this book is about. Many well designed and well engineered products fail every year in the marketplace while crudely designed and engineered products succeed. Your customers want, and pay for, the functional RESULTS from your product, not great industrial design and well executed engineering. Yes those may be important as you get deeper penetration in your target market but my belief is you shouldn't overly worry about them up-front.

Whether you work with a design firm or the in-house design/engineering staff at a manufacturer, be clear what you want them to do and what you are willing to pay for at this stage. Yes, you want something that looks like a finished product in the end but it doesn't have to be the ultimate in modern industrial design either. DO make sure though that part manufacturing and assembly expenses are considered during the design and engineering because those will make a real difference in your chances for success.

My bias is to go for the in-house design/engineering capabilities of a manufacturer that you can work with. They will very likely cut you some slack on the price if they are sure you will be ordering production runs from them. One way they will guarantee this, of course, is to own the electronic version of the design they do for you. Make your own decision depending on the complexity of your product, the price they charge for the work, and your belief in a win-win attitude on their production pricing.

For a product that takes $30,000 in design effort but for which you are really charged $10,000 or $15,000, they should "own" the electronic file for a while but for a product where the design effort is $1,500 and you pay it all you should certainly feel entitled to a copy of the file. I think manufacturers that plan to hold you over a barrel by keeping the electronic file are extremely short-sighted because the antagonistic kickback from inventors is to quickly move production overseas even if it requires paying again for design and engineering.

If you are a manufacturer reading this and don't know the story about the contest between the Northwind and the Sun, now might be a good time to read or listen to it. True, there will be some people that blatantly take advantage of generosity but you, as a manufacturer, will likely make more from goodwill, repeats, and referrals by not overly concerning yourself with the few who take unfair advantage of you. If you are an inventor reading this and you want to see respect for all inventors, you shouldn't plan to be among that small minority.

If you do go to a design/engineering firm they may very well want to run you through their standard 7 (or whatever) iteration design process. You are a business person, work out with them what you can and will pay for. Smaller firms that don't cater to the industrial giants are your best bet. Search the Internet or start by calling "Designers-Industrial" from your phone book yellow pages. If a firm you call or contact doesn't work in the area you need, ALWAYS ask them if they can suggest a firm or individual that does.

Design References List

Some recommended reading materials on engineering, design, and manufacturing are:

Product Design for Manufacturing and Assembly, Boothroyd & Dewhurst, 1994 (earlier, similar books should be useful also).

Designing for Manufacture and Assembly, Therese R. Welter, Industry Week, September 4, 1989.

Designing for Simplicity, Byoung Sung Kim, Mechanical Engineering Design, Nov. 1999, pp. 34-36.

Effective Product Design & Development, Stephen R. Rosenthal, Business One Irwin, 1992.

The House of Quality, John R Hauser and Don Clausing, Harvard Business Review, May-June 1988, pp. 63-73.

Juran on Leadership for Quality, J. M. Juran, The Free Press, 1989.

Familiarize a Patent Attorney with the Prototype

Once you have a fairly solid working prototype and are reasonably certain no more "gotchas" are going to have to be dealt with you have something a patent attorney can accurately yet broadly describe. Remember, you get a patent on what is described, not your revisions to make it work if you gave your original idea to a patent attorney. While you may want to interview a few patent attorneys before choosing one I suggest you pick one, then agree to pay the rate for CONSULTATION only to start with.

Remember, if it is free advice you want, you will get exactly what you pay for. It is amazing that honest patent attorneys stay in business. Many things are not patentable because there is nothing "new" to them and the attorneys are well aware of this. That is one reason the "FREE" advice from honest patent attorneys will often be "No, that is not patentable (as I understand it and without considering exceptions I can't think of just now)." Since you have winnowed down your first hundred ideas to this one you are presenting to your attorney, your invention is much more likely than most to get a patent that will be <u>worth something</u>.

Your patent attorney will probably insist on their own patent search. This is probably a good idea even if you did a search in an earlier step. It is almost mandatory if your final invention is significantly different from the original concept that you did your patent search on. Particularly where competing products already exist, you will also now have exact details from your "final" prototype that can be checked against existing patents.

Patent Search Costs

While I haven't experienced it, I've heard stories about legal firms charging $1,000 to $2,000 for a patent search then paying a paralegal near the Patent Office in Arlington, VA $250 to do a U.S. Patent Office only search. The rest of the fee is for the "opinion." I've paid $750 (which I think is reasonable) for a U.S. Prior Art search that includes an online literature search and a search at the USPTO by the attorney I paid. Ask about what you are getting for your fee and who does the search and how.

At a superficial level you can often tell what prior patents will conflict with yours but you really should pay for a professional patent attorney's (or agent's) written opinion. The real patent is for the claims, not all the rest of that easy stuff that you can read and understand. The attorney can read the claims and compare them against what they expect your claims might be and give you a probability estimate of your getting a patent with the significant claims for your invention intact.

If the search finds significant conflicts that are not expired you may want to design around them or try to license rights to them from their owner. Remember, if you are finding conflicts now and you find it fairly easy to design around them, your competitors very well may have the same opportunity. Evaluate carefully—and try to make a business decision rather than an emotional one about the money you invested in the prototype.

A properly written broad claim for an aspect of a novel invention can be very difficult to engineer around because of the "Doctrine of Equivalence." Essentially that means that if the elements of a new invention serve the same function as a patent's claims even though they do not literally infringe, then the new invention infringes anyway. That's the way you want it for your patent and that's the way you should want it for everyone else's also. Examples are a hole is a hole, it can be any shape, and a bar is a bar even if it's shaped like an ess. If the shape has a specific additional function that is not covered by the original claim, then you might be able to make a specific narrow claim but that does not make it such that you do not infringe the original patent if it is still in force.

You also want to explore design versus utility patents if you don't understand that aspect of patent law. Design patents are often not worth much

---

 Engineering around your competitor's patent is easy.

---

but there are exceptions. The best example I've heard of was for patterned women's hose where the company coming up with the original idea got design patents on many different designs then used those patents to extract royalties from competitors wanting to do patterned hose.

Specifically ask the attorney what it will take for you and them to go through the patenting process. If they give you a fuzzy answer about number of hours that may be the best you'll get, but at least they should give you some separate ball park ranges for a patent search and for preparing for filing plus the government fees.

Be aware that you will probably be charged separately for, and may have to have to pay in advance for, one each of Patent Search (and Opinion), prepared Provisional Patent Application (or its review if you choose that route and prepare your own), Patent Application and fees, Prosecution of the Application (i.e., reworks to get the examiner to accept it), issue fees, and Possible Infringement Review.

Expect a total of $6,000 to $10,000 (including government fees and assuming no interference and minimal prosecution issues) for a simple invention even if the attorney says they will <u>prepare the Application</u> for $2,500. Don't charge into an application yet, just get a feel for what goes on AND FIND A PATENT ATTORNEY YOU TRUST TO WORK IN YOUR BEST INTERESTS. If you think you will be going for foreign patents as well as U.S. you should ask about their experience and capabilities in the foreign patent process too and choose someone you are comfortable with.

Take Charge of the Patent Process

<u>You are in charge of the patent process.</u> Listen to your patent practitioner for advice but make your own decisions and make it absolutely clear you will pay for what you ask for. In the end the Patent Office can still reject your patent regardless of what you have invested in it. Your attorney (and your money) will not control the outcome—and they wouldn't even if you were a big corporation.

Robert Merrick cites as an example an invention for which he asked his patent attorney to skip an initial patent search and go directly to preparing a patent application despite the attorney's initial assertion that the invention was NOT patentable. The invention turned out to be patentable after all and Mr. Merrick made a handsome profit from it. HOWEVER, that was an exception—if you are counting on that good luck you'll likely go broke first.

A friend I met at our local inventors club said he had trouble with his first attorney who said his invention wasn't patentable. He went to another attorney and did get a patent. He also bought an inventory and got some marketing partners and 3 years later is still in the hole financially. He figures if he can get out of his agreement with his original marketers and get a new marketer to sell the first production run that is still in inventory he'll be on his way. I have not had a chance to analyze his product or market but my guess is that he has a novel product all right—but it is not as good (in the BUYER'S eyes) as competitive products or BFH solutions even though a few have sold.

Not Patentable? So What!

If you get a patent in the future, of course you have a good marketing edge that will certainly help you achieve the projected profits you computed in earlier steps—and, of course, still assuming that the market WANTS your product. **But what if you won't be able to get a patent? Or what if a patent could be easily engineered around? If your initial market analysis**

and cost projections done in earlier steps are still valid then your answer should be "So what!"

If you still stand to make a profit, and Mr. Merrick and Mr. Tripp both cite examples of unpatented product ideas that do, there is no reason you shouldn't succeed. All a patent does is provide a temporary roadblock to

---

 You can't get a patent and the market is likely to attract competitors or you can't compete on price.

---

knockoff competition anyway—it does not block competition from others with different embodiments of solutions to the same problem. Do you want the "Inventor:" designation on an approved patent or do you want the money?

You must make your own decision but as a marketer I have already made mine—I'll only help inventors who are interested in making money by providing products the market wants and that can beat competitive products. If it's done right, just being first to market with a product that sells can discourage your potential competitors.

A big caveat to the "So what!" point above is that, without the patent, you will be vulnerable to near-exact knockoffs. This should worry you significantly only when your market is huge enough to be attractive to someone who only gets say 5 or 10 percent of it or when you will find it virtually impossible to be the low-cost producer over the long term. This is where having a simple product but relying on subcontractor-manufacturing (with the required double profit mark ups, one for the subcontractor and one for you) can kill your product's attractiveness to consumers and retailers. In other words you need to understand your customers' price sensitivity and be able to pare costs to the bone after your product is off and running if knockoffs can be easily introduced.

On the other hand, even a good broad patent isn't a guarantee. Take the case of the television camera/picture tube (Image Dissector/Image Oscillite). The patents **(note the plural)** were solid and broad and essentially guaranteed that anyone wanting to make and market what we now call televisions, or to electronically capture and send video images to them, had to license rights to do so.

Unfortunately the market for TVs didn't become huge till after the patents expired. Fortunately the technology was complex enough to lend itself to many novel and unobvious patentable improvements. Unfortunately the original inventor didn't invent all the improvements but he did invent over 100 of them and was still a smashing financial success for the more than 15 years he put into the television development effort. Television, by the way, while instantly and wildly popular, got a fairly slow start because a TV is useless by itself. The infrastructure of Broadcasters and News/Entertainment Programming also had to be created to support consumer TVs.

Patent or not, you can and should provide yourself with trademark and copyright protections. A good trademark can be a pile of gold and yet be very inexpensive to get and defend (more later).

Remember, at this point all we have is a working prototype, we still haven't proven it will sell, so all money spent so far, including patent or consultation expenses incurred, are still at risk. You can still keep your financial risk down by not yet submitting a patent application so you might want to talk the pros and cons of that over with your patent attorney before proceeding with the next crucial marketing step.

 15. Your solution to a restatement of the problem is much simpler and likely to be "better" to the consumer. Restart Step 0!

16. You can't design and make a prototype that works satisfactorily.

 EXCEPT with an extra gizmo or product that is necessary for efficient/effective use.

17. No prototype maker or manufacturer will provide a quote or offer you a cost-plus development agreement.

 18. Your "fixes" to unanticipated development problems require more complexity or exotic materials that boost your profitable sale price above a reasonable level.

19. You need university research expertise but are unwilling to put up with the grant or other financing hoops.

Explanations and scoring suggestions:

15. <u>Your solution to a restatement of the problem is much simpler and likely to be "better" to the consumer.</u> Sometimes solutions from all perspectives (restatements) sell as attested by the "getting business cards into Rolodexes" problem. Make sure your solution has what a significant part of the buyers will perceive as an edge if there are other solutions on the market. If there are no competing solutions on the market, what restatement solution do you and your snicker testers believe will be best in the long run?

 1%—3 or more restatements did not lead to any solutions you or your snicker testers concluded were better.

 5%—You worked hard at it and couldn't come up with any restatements of the problem.

 50%—A solution to a restatement looks better from a customer perspective. (Suggestion: take the new solution back to STEP 0.)

 75%—Your restatement solution is easier, cheaper, better, but it doesn't look like there is any chance at intellectual property protection.

 75%—You believe restating the problem is a worthless intellectual exercise and didn't do it (even though it could net you three solid patents and total ownership of the market).

 90%—Your invention works superbly but the user has to have an extra item that is expensive or inconvenient. A shirt "button" that requires a fasten/unfasten tool probably won't catch on.

99%—Your restatement turned up an existing solution that is better than your original invention and it currently is satisfying the market.

16. <u>You can't design and make a prototype that works satisfactorily.</u> If the problem your product solves is a significant one and your invention does work in some instances then it may be acceptable to narrow the scope of your problem statement. Be very careful stating your problem though. It is NOT unusual for product ideas to bite the dust at this stage. In fact, if you found competing patents, but their products were NOT on the market, it's a safe bet that better than 50% of them don't work satisfactorily. If you go ahead with sales you will probably find you get initial sales but they mysteriously "dry up" as if "the word got around."

 0%—The invention works in all reasonable situations and even under adverse conditions.

 25%—The invention works in a few specific, but significant, situations.

 50%—The invention works in a few specific, but minor, situations.

 95%—The prototype doesn't quite work reliably but you expect that problem to not exist in production line versions.

 100%—The invention doesn't really work in any practical situation.

17. <u>No prototype maker or manufacturer will provide a quote or offer you a cost-plus development agreement.</u> This may be a good clue that your choice of materials won't match the use you are trying to put them to. It also might be that your engineering or design is far from industry norms. ASK FOR the reasons you get no-bids—you'll likely learn something valuable.

 0%—You got bids lower than you expected from several manufacturers.

 5%—You got a reasonable bid from one manufacturer but several others turned you down.

30%—You got only one offer and it was on a cost-plus basis but you have seen other successful products on the market that require your kind of manufacturing. (Suggestion: see if there is a slightly different type of manufacturer that you should approach.)

 50%—You got no bids but one or more will consider a cost-plus prototype development arrangement.

 99%—You get no bids and they all say what you want is either impossible or impractical with known technology.

18. <u>Your "fixes" to unanticipated development problems require more complexity or exotic materials that boost your profitable sale price above a reasonable level.</u> It is very easy to fall into the trap of "I'll make this work before I die!" Try to separate "you" from "your invention." "IT" has no animosity or any other feelings toward you and is not being cantankerous. Are you outside your field of expertise?

 0%—Your invention required no or only minor adjustments to make it work as you expected.

 10%—Your fixes are not too major. The expected boost in production cost is less than 20%.

 30%—Necessary fixes probably boost your production cost by 50% but there are no real competitors.

 60%—In a field with competitors, your production costs will easily make you the high price producer but with a definitively superior product.

 75%—Your production cost will make you about equal to competitors and you have no significant extra benefits for your buyers.

 99%—Your production cost will make you significantly more expensive than competitors and you have no significant extra benefits for your buyers.

19. <u>You need university research expertise but are unwilling to put up with the grant or other financing hoops.</u> Research is rarely done quickly either so your next idea might be a better one to do first.

 0%—The invention is totally producible by many manufacturers with well known materials and technologies.

 40%—You are willing to do all the required grant paperwork and are fairly certain that the expertise to solve the problem is available.

 75%—Government bureaucrats are not your favorite people but you'll do the paperwork to see if you can get the money for a "cooperative" expert you know.

 95%—An expert "grant writer" appeared from nowhere and for $5,000 up front (and 40% to "administer" the grant) they will write and submit a grant proposal.

 99%—Most experts you talk to when you start exploring grants are pretty certain that what you want can't be done.

 100%—You don't want anything to do with grants and besides that you are not certain what you want can be done.

STEP 5—Sell a Few of Your Product.

Patent considerations for this step: Up to this point you should have been diligently getting non-disclosure agreements signed and been careful to keep your dealings explicitly confidential even when you don't get a non-disclosure signed.

For my money, I'll prepare my own Provisional Patent Application and file it myself with the $75 fee. I will, however, incur the trouble and expense of having my patent attorney review the provisional application, and I will adjust it as necessary, before I send it in. Filing a provisional application starts a 1 year clock ticking in the U.S. (and many foreign countries). You have 1 year in which to submit the real (non-provisional) patent application. Your filing does 3 things:

1) it (optionally) establishes a priority date,

2) it generally allows you to make public your invention without losing any rights to it in foreign countries, and,

3) it lets you put "Patent Pending" on your product and/or marketing materials.

Your provisional application must be a balanced amalgam of thorough and broad and specific but need not detail patent claims or prior art. If you have a really truly working prototype you should be able to do a pretty good job drawing up the provisional application. The key is that your future real application CANNOT, if you want the most certain U.S. and foreign patent protection, add new features, concepts, etc., that DO NOT HAVE SUPPORT in your provisional application.

Patent Application Terminology

"What's in a name?" The Provisional Patent Application (PPA) is formally called the Provisional Application for Patent (PAP). The real patent application (RPA) is also called the formal patent application (FPA) and its official name is probably just Application for Patent though it is semi-officially also called the Nonprovisional Application for Patent and Nonprovisional Patent Application (NPA). Ready for the test? "A rose by any other name would smell as _____."

(Actually, it can add stuff, but the added stuff will have the priority date of the real filing and, if your public disclosures included the added stuff, you've probably lost any foreign filing rights to the new stuff.)

That is why it is good to have your attorney understand your invention and review your provisional application even if it increases your final legal costs by 300 or 400 or so dollars. Actually, patent attorneys can diddle your patent claims for you even a couple of years after the patent is granted if you pay them money for it. The catch for you is you must decide if such diddling is economically useful or not. (Mostly, it's not.)

You are better off over describing the details of your invention and referencing the specific parts of the drawings and including details and

options than you are under describing. Just because the fee is cheap doesn't mean the provisional application filing can safely be sloppy. The drawings for a Provisional Patent Application can be informal but must show everything and be understandable (perhaps eventually, but much less than 1% of the time, to a judge or jury).

It doesn't matter if you accidentally describe prior art but you do not have to identify prior art in the provisional application even if you know

 You file a Provisional (or full) Patent Application without being certain that it is complete.

about it. Do not trust a patent practitioner that tries to convince you otherwise, they make money on the extra rigamarole but the Patent Office doesn't care.

You may want to include a claim, or at least what looks like a claim, in your provisional but I won't bother to. Some attorneys argue that some foreign countries <u>may</u> not accept the PPA filing as a legitimate patent filing because there are no claims in it.

Note that "<u>may</u>" is a scare word. There is no case result that supports their argument yet and the director of the WIPO has

Fear Words

There is an advertising principle called "FUD." It stands for Fear, Uncertainty, and Doubt. It is a last ditch tactic that, I believe, should only be used with Fear and then only when the purchase truly is in the best interests of the buyer and they are not readily comprehending rational, correct, and appropriate arguments. When used by patent professionals it is very UGLY.

Whenever your "advisors" resort to raising doubts, uncertainties, and fears in your mind, (a common tactic of scam artists) get an independent advisor that doesn't stand to make or miss out on money based on your decision. You need to make decisions based on realities and facts.

indicated that PPA filings will count with regard to subsequent PCT patent applications. The attorneys counter that the PCT countries are sovereign states and not bound by decisions of the WIPO director. Do you want to concern yourself with selling your product (or paying for a patent) in a country who's leadership advocates theft? Not me, I won't worry about putting a claim in my PPA.

If you start selling your product, or even just offering it for sale, or provide it for public use (rather than explicit, less than obvious test use), or announce it to the public or have a research paper published on it BEFORE you file your PPA you will not be able to get foreign patents and you will have started a one year clock in the U.S. In the U.S. you must file the full patent application or a provisional application before that year runs out or you will lose any rights to a U.S. patent too.

The TRICK. Yes, the above paragraph is correct. You can protect your U.S. rights to a patent by filing a Provisional Application for Patent on your invention up to one year after public disclosure. That will give you another year in which to file the full application. Let's see, that means you will have lost any foreign rights (but the U.S. market is so huge anyway) and you can postpone the expense of a patent application for about 1 year and 9 months (you have to give a patent professional time to write it). Wow, that's great!

But! If it takes another 2 years to get the patent and since you cannot stop others from making, selling, and using your invention until after a patent is granted, you may just have given (perhaps less than ethical) competitors about a 3.5 year window in which to exploit your invention. In many cases that may exceed the lifetime of the product. Fad products rarely gain much benefit from patents.

Son of TRICK. Even if you file a provisional application just before your first sale as I suggest, if you find that the market demands significant changes (probably because you didn't do your homework in the first place) that mean your provisional application does not adequately SUPPORT what will be your full application claims, you can abandon your initial provisional application (at the expense of losing any foreign rights) and file another provisional application on the same invention that will still give you the second year after public disclosure in which to file the full application.

Yes, the U.S. patent laws are much more inventor friendly than foreign patent laws, but playing games with your patent filing just because you can is not a good plan. It starts opening your patent up to various litigation validity attacks and it will usually cost you more in the long run. Besides, the real rewards don't come from getting a patent (although 98 out of 100 inventors might, for their own ego's sake, want to convince you otherwise), the rewards come from the marketplace where people pay real money to have your product.

Provisional Application for Patent Misconceptions

A couple of misconceptions about Provisional Applications for Patent should be cleared up here. First the provisional application's content is NOT examined by the USPTO but you may hear of their being "rejected" from time to time. It is not that the content or application is truly rejected, rather the USPTO can require that you fix the application to meet their "technical" requirements including paper size, page margins, dark ink, page numbering, etc. as noted in the sample on the next page.

These technical requirements are spelled out in 37 CFR 1.52. Yeah, right! There is a link to Title 37 of the Code of Federal Regulations in the lower left corner of the "Patents" page off of the USPTO home page (www. uspto.gov). This and the Manual of Patent Examining Procedure (MPEP) accessible from the same place are pretty dense reading so a current version of the Nolo Press books may be easier. The rules for PAPs are more lenient than those for FPAs.

You should also be aware that the rules really stretch the law's requirements of understandability and simplifying the examining process. Some of the rules arose from questionable pre 1985 court decisions and the USPTO is slowly relaxing them back to a more reasonable level.

Since the Provisional Application for Patent is not examined it cannot mature into a granted patent and it used to be that a completely new full patent application had to be filed. I still recommend having an attorney prepare and file a completely new full patent application at the appropriate time but it is now possible for you to have (with the right rigamarole of course) the Patent Office "convert" your PAP into an FPA. The big catch, of course, is that for a PAP to be properly "convertible" it must meet (or be brought up to) all the standards of a full patent application. As I see it, the new "conversion" rule just gives patent attorneys one more way to sucker naive inventors into paying for a full patent application first.

You also may read that if the 12 month deadline date for filing your full application referencing a PAP falls on a weekend or holiday that you MUST get your FPA in BEFORE the weekend or holiday. That is now (effective November 29, 1999) no longer the case, you can get it in the day after the weekend or holiday. If you are cutting it that close I reserve the right to consider you a nutcase anyway. Get the FPA in the week before the deadline and do it via a service that gives you a verified date of delivery.

Lastly, Congress in its great wisdom, has added a new "confuser" to the pile and created something called "provisional rights." These rights have nothing to

CFR "Technical" Requirements Sample

(a) The application, any amendments or corrections thereto, and the oath or declaration must be in the English language except as provided for in §§ 1.69 and paragraph (d) of this section, or be accompanied by a translation of the application and a translation of any corrections or amendments into the English language together with a statement that the translation is accurate. All papers which are to become a part of the permanent records of the Patent and Trademark Office must be legibly written either by a typewriter or mechanical printer in permanent dark ink or its equivalent in portrait orientation on flexible, strong, smooth, non-shiny, durable, and white paper. All of the application papers must be presented in a form having sufficient clarity and contrast between the paper and the writing thereon to permit the direct reproduction of readily legible copies in any number by use of photographic, electrostatic, photo-offset, and microfilming processes and electronic reproduction by use of digital imaging and optical character recognition. If the papers are not of the required quality, substitute typewritten or mechanically printed papers of suitable quality will be required. See §§ 1.125 for filing substitute typewritten or mechanically printed papers constituting a substitute specification when required by the Office.

(b) Except for drawings, the application papers (specification, including claims, abstract, oath or declaration, and papers as provided for in this part and also papers subsequently filed, must have each page plainly written on only one side of a sheet of paper, with the claim or claims commencing on a separate sheet and the abstract commencing on a separate sheet. See @§§ 1.72(b) and 1.75(h). The sheets of paper must be the same size and either 21.0 cm. by 29.7 cm. (DIN size A4) or 21.6 cm. by 27.9 cm. (8 1/2 by 11 inches). Each sheet must include a top margin of at least 2.0 cm. (3/4 inch), a left side margin of at least 2.5 cm. (1 inch), a right side margin of at least 2.0 cm. (3/4 inch), and a bottom margin of at least 2.0 cm. (3/4 inch), and no holes should be made in the sheets as submitted. The lines of the specification, and any amendments to the specification, must be 1 1/2 or double spaced. The pages of the specification including claims and abstract must be numbered consecutively, starting with 1, the numbers being centrally located above or preferably, below, the text. See §§ 1.84 for drawings.

do with Provisional Applications for Patent but instead are related to full patent applications where you express an intent to file for foreign patents. Normally this is done with a simultaneous PCT application (discussed later) and the PCT application, to conform to international standards, will be published 18 months from filing.

The "provisional rights" that you get are that AFTER your patent is granted you may, at your own trouble, collect a "reasonable royalty" (but NOT lost profits) from anyone "infringing" your patent between the PCT 18 month publishing and patent grant. One catch to these rights is that you can only recover "royalties" after the infringer is "actually notified" of your patenting activity. That does not mean you have to know about their infringing and personally notify them. If they subscribe (or have others subscribe for them) to a service that publishes or republishes these 18 month information releases, or your products (under PAP rules) are marked "patent pending" and they copy them, you will likely be able to recover "reasonable royalties." Only if they "innocently" reinvent your invention and you don't find out and notify them are you likely to be out of luck. In the unlikely event you get into this situation I would suggest that you try to negotiate a license agreement with them first rather than threatening or starting court action.

In a nutshell, my advice is: have your product development done, file an attorney reviewed PAP that fully describes the product but has no claims or prior art, and get the product on the market ASAP after that.

Make a Real Batch

Depending on cost and quality issues you need to have a test quantity of your product made for sale, i.e., test marketing. Big companies do test marketing all the time. They often do it on a city-wide scale in a small to medium city whereas you may only do it in two to ten stores (or by other means). The test quantity can either be an initial <u>short</u> production run or a batch of prototypes that are extremely close to the real thing.

If prototypes are significantly (in the BUYER'S perception) higher or lower in quality than production devices then a market test with prototypes will NOT provide valid results. The worst situation is that you use prototypes that are significantly better in quality and they sell really well thereby fooling you into believing you can succeed with a really long first production run in order to maximize your profits.

You should also prepare some professional looking packaging—but don't overspend trying to make it your production packaging yet and don't worry if it (or your prototypes for that matter) boosts your test market product costs well over your asking price per product. Your objective here is not to make a profit—it is to slash your risk, and the perceived risk of retailers, etc., that will possibly handle your product in the future.

If tooling is required to make your product, and it usually is for plastic or metal parts, I recommend that you go with the most inexpensive tooling possible that is consistent with your quality requirements and the size of the prototype or test production run you need. In the long run, for this particular product, should it be proven to sell, you may actually incur higher expenses than if you went directly to production tooling. In other words you'll pay for both the inexpensive prototype tooling and the much more costly production tooling.

For example I got a quote for a "finger and ring measuring caliper" requiring 4 opaque plastic parts, one clear plastic part, and printing. The engineering and design and hot stamp fixturing (for the printing) came to $3,450. Production tooling (single family mold) totaled $15,100 and prototype tooling was only $2,600. Production items were priced at $1.32 each for a minimum run of 5,000 and prototypes at $4.00 each with a 100 minimum.

If my (manufacturer's) projected selling price is $1.50 each, how much will I make on the prototypes? I'll be in the hole $6,300 if I sell them all at my desired price. If I go with a full minimum production run and tooling and sell them all at my price I'll be in the hole a whopping $17,550. If they don't sell at all I would lose only $6,450 on the prototype version but $25,050 on the production version. (In either case I'll lose everything I've spent on patenting also.) If they do turn out to sell I'll wind up losing only $2,850 of the money I invested in the prototype run.

On the other hand I may discover they sell so well that I want to go directly to a 5-family mold for production at a cost of $50,000 or so and skip the $15,100 of the initial single family mold. That will likely reduce my per unit cost to $0.80 or so but require a minimum run of 25,000. Since I'll know they'll sell before taking that route, the "wasted" $2,850 when I put the

$6,450 at risk on the prototype mold will be completely recovered by sales of only the first 5,481 of my first minimum production run.

Inexpensive vs Expensive Tooling Calculations

Confirm my math above as a homework practice exercise. Figure out how much farther I will be ahead after sales of 25,000 using the 5-family mold versus using 5 iterations of the single family mold I would have bought to avoid "wasting" the $2,850 on prototype mold costs.

Packaging

Your test packaging should include your company (or DBA) name and logo and a (potential) Trademark and possibly even a copyright notice. Oh no! More lawyers and money. Keep it simple at this point. Trademarks do NOT have to be registered before they are used. In fact your life will probably be simpler if you use your trademark before registering it. Also, a lawyer can be helpful but is not required to register a trademark. Think up a number of possible trademarks.

The best trademarks have both no meaning and an obvious meaning (e.g., Kleenex®, Realtor®, Plax® come to mind). They are made up words that are obvious fits when you know what they are for and are almost understandable when you don't. Kleenex, by the way, was not invented for nose-blowing, it was invented for wiping off makeup with coldcream. Marketing repositioned it later. Made up word trademarks are hard to invent especially since misspellings of real words don't count as made up words in Trademark law. For example "Klub" is considered equivalent to "club," "Kurly-Q" to "curlicue," etc.

Trademarks should not usually remind your buyer of your competitors either although that might help identify what your product is when you are new (remember Pepsi-Cola, a trademark dating back to at least 2/15/1897 and now almost invisible due to the use of only the Pepsi trademark in order to dissociate from a famous trademark of 6/28/1887). Follow your trademark with the ™ superscript to let it be known that you are claiming a common-law trademark. Yes, this is 100% legal, you don't have to get permission from anybody. See Chapter 7 for more details on trademarks, trademark searching, and copyrights.

Your trademark, however, should NOT receive the most emphasis on the packaging. The biggest and easiest to spot words on your packaging should clearly state what the product is or does. "Bicycle storage rack," "Pain relief," etc. get right to what the customer might want. The next most important part of the text on your packaging tells the customer the benefits they will get

 You launch your product sales without a common-law trademark.

from your product (and if space permits, the product features that provide those benefits). The list should generally be from most important to least important with perhaps some raising of the key benefits of your product over competitors products.

Also on your packaging show a picture or drawing of the product in use as well as the steps to use it, etc. I could write a book (and other people have) on packaging but your best bet is to go shopping, find products with packaging you think is effective and useful to the consumer (not necessarily pretty or eye-catching) and emulate it. If you find your type of product pretty much exclusively has a particular type of packaging be sure you use that type of packaging too. You want to keep your retailer happy and keep their, and their clerk's, lives simple.

But don't adopt the "trade dress" of some other manufacturer as part of your packaging or you could get in trouble. I wouldn't put Jim's-Cola in a Coke style bottle or Jim's Film in a yellow box or do the current "hot" thing of cheap water softener salt sellers and copy Morton's yellow bag. If you have to "sort of" deceive someone to get them to buy your product are you all that sure you want to be selling that particular product anyway?

It is, however, sometimes important to differentiate your product through packaging (think L'EGGS pantyhose) but be sure your channel will accept that. In the case of L'eggs, the manufacturer initially did the stocking (no pun intended?) in their own custom racks in the store, in other words, all handling except at the cash register so the retailer and their stocking clerks didn't have to do anything different.

Patent Pending

If you filed a Provisional Patent Application or a full Patent Application do not forget to put "Patent Pending" on your product (if possible) and on your packaging, marketing materials, instructions, etc., for sure. If you consistently, as opposed to accidentally, fail to do this, in the worst case, the courts can construe that as a desire to make the technology available freely for "the benefit of mankind" and in the best case, the courts will simply make it impossible for you to collect damages from infringers prior to proper notification and the opportunity to gracefully exit.

At this stage the patent pending notice should serve to discourage copiers but, until a patent is granted you will have no rights to exclude others from making your invention. If sales are going gang-busters when you do get a patent you will likely have significant quantities of "patent pending" marked products in the distribution channel at that time. You will, of course, want to start properly notifying the marketplace of your patent number as quickly as possible after you get it but it is unrealistic to recall and mark the packages already in the pipeline. A company "innocently" infringing one of those can be requested to stop but you are unlikely to collect damages. If they have some market share and good distribution channels you may want to see if they are interested in licensing before you threaten them with legal action.

Benjamin Franklin, for example, intentionally made all of his inventions available freely for the benefit of mankind. He was also quite chagrined when copycat manufacturers copied only the look of his stove without understanding or duplicating the technical details inherent in the principles of its efficient operation. Fortunately for Ben, his reputation didn't rest solely on the shortcomings of the inferior copycat products.

Do not put "Patent Pending" on your product or materials if you have only filed a Disclosure Document or have taken no patenting steps with the USPTO. You will be in violation of the law and court time, lawyers' fees, penalties, etc., are not what you need at this point in your venture.

Liability Insurance

If you see a reasonable possibility for your product to cause injury that might make you liable for damages you may want to have a discussion with a business insurance agent. While you won't have much leverage initially, be aware that business insurance rates are negotiable. You often don't have to just accept what the agent tells you. If you can make a case for a lower rate do it! On the other hand I wouldn't recommend that you set your deductible so high or the insurer's cap so low that you'll be bankrupted by the first serious claim.

At the market test stage you are not likely to be able to negotiate much because there is no data for your product although an insurer may have data on similar products. For most products liability insurance should be fairly cheap. The hand tool that my friend couldn't get his first attorney to prepare a patent for requires an annual payment of $2,000. Given that the first batch is mostly still in inventory that could probably even be negotiated down.

Be careful when looking at the risks of your product. I worked on one project where the inventor thought his product could pinch a finger at its worst. True, the product itself could pinch a finger, but the real liability lay with the product in use. If the product in use FAILED it could drop debris that could easily result in a fatal accident on public roadways. A good insurance agent will be able to spot those kinds of risks. A good insurance agent may also be able to suggest appropriate warnings and instruction disclaimers for your product.

Market Test

You will find out two things with a market test if you use products from a production run. Does the production process work and provide the level of quality you require (it usually never hurts to exceed the expectations of your buyers), and do the production products sell? If it is a prototype test you will only know whether nearly similar production products will sell.

You've no doubt seen (or been accosted by) those "taste a sample" pushers in grocery stores. You can do a similar thing yourself in an appropriate store.

 Your sales test products are hand finessed quality prototypes that actual production won't match.

Pick two or three locally owned stores if possible that would best carry your product. With the permission of the store owner or manager, and an agreement on how the proceeds are to be split (I suggest starting negotiation with an offer of a 50/50 split of receipts but be prepared to go to 0/100 in the store's favor), set up a very small stand which you (or your (paid) marketing representative) will man.

Your best bet is to pick non-chain stores that are owner operated but you can get into at least regional chain stores with the right product AND if you approach the right person. Trust me on this, if every store owner you show your product to tells you it is store policy not to do test marketing or that the chain's main office must approve it, what you should be hearing is they don't like your invention. Their "policy" is just a polite way of saying that. (Caution: in small stores that are part of regional chains it is often possible to talk to the "manager" who is not a manager at all, just a clerk with a big title. Rarely can these people make a decision or help you except to give you a contact name at the main office.)

Always approach the store professionally. Wear professional attire, have a quality product and attractive packaging, and a business card. That said, I once ducked into a store on a rainy Saturday in my damp blue jeans with my hair a mess and with no business cards, told the owner I was the marketer for a new product and asked if they would be interested. They said send some info. The next business day the info, on letterhead, and a business card were in the mail to them. They accepted the opportunity.

Have spaces for 20 or 30 (pick an appropriate number) of your wares but only fill about 90% of them (you don't want anybody to perceive they might be the first at this point). When someone comes along try to make eye contact and ask if they would mind helping you answer a few questions. Explain you are a marketer working for whatever your company (or DBA) name is and you have a new product that you are just introducing. A clip board, a pad of questionnaires, and professional attire will all be useful in getting the confidence of prospects. If your product is particularly complex a demo video showing it in action might also be appropriate.

You need a prepared introduction to what the product is, what it does, why it is new, what it does better than competitive products, what its price is (in future tests you may want to ask prospects to estimate its value but not yet), and ONLY 1 major benefit the customer could get out of the product—fit all that in 30 to 45 seconds before you hit them with your first question. You come up with the benefit list in advance but only burden each prospect with 1 benefit and note which on the questionnaire you complete for them.

Ask a few specific yes/no questions, multiple value (better than x—strongly agree to strongly disagree) questions, and open ended (quality perception, design, safety, etc.) questions while trying not to tie up more than 2 or 3 more minutes of their time. Ask questions that you genuinely want answers to—except "Will you buy it?" Some examples: Does the X problem affect you? Would your selection of such a product be style or color sensitive? What do you currently do to handle problem X? If, during your discussion, the prospect mentions other benefits or (gasp!) failures, write them down on their questionnaire whether you already know of them or not (you're gathering statistics remember).

Conclude your questions with "Would you find a product like this useful?" If they say no, thank them for their time. If they say yes, ask them if they would like to buy one today. If they say yes hand them one and a $2.00 (or as appropriate) off coupon, good only on the product, to turn in at the cash register. Thank them for their help and for buying and hand them an envelope, or have one already inserted in the package, (self addressed to you and postage paid) with a user questionnaire to fill out after trying the product.

While people won't be coming to the store specifically to buy your product because they know about it and want it, the problem of public awareness is in part solved by your (polite) in-your-face stand. You can also do entryway and window posters (good quality—probably paid for—not hand lettered in multiple marker colors) that give some advance warning of your presence and may arouse interest.

If you can get the store owner to believe in your product and they routinely do newspaper inserts or newsletters or other regular advertising that they are financially committed to anyway, you might be able to get them to advertise your product at no charge along with others in one of these media. The major catch here is that there may be as much as 2 months lead time in getting your product announcement/ad ready and your actually setting up the stand. You must also factor that advertising into your evaluation of the sales level.

Price Testing

I would recommend trying a range of prices at different stores also (e.g., $5.95, $6.95, ..., $9.59 or $19.95, $24.95, $29.95) to see what the result is. Be sure the stores are far enough apart to generally get different shoppers (that might be 20 miles in a rural area but 5 miles in an urban area). Be cautious here though and ask for each store's gross annual sales so you will have a crude way to determine whether the difference in sales volumes between stores is due to differences in traffic or your offer price. Most managers/owners will willingly tell you gross annual sales but if they won't you may be able to get it from your state's sales tax office or from Dun & Bradstreet (for a fee, currently $20 via credit card and you must register) at www.dnb.com or other business info sources.

You will leave a rack or box or set of product at the store for them to sell in your absence also. Stapled or otherwise attached to each sale unit package will be a proof of sale tag that is meaningless and reasonably unobtrusive to the customer but which store clerks will be instructed to remove and put in the cash register. (Big manufacturers often include the UPC (Universal Product Code) barcode as part of this tag but you don't have one yet. How to get one is covered later.) The purpose of the coupons is not just to thank the customer. The coupons and the proof of sale tags combined provide the retailer, in their cash register, with tangible proof that your product sells—and (hopefully) not just when you are there. (Packing the cash register by sending shills to do the buying could be construed as fraud if you use that number of sales to convince sellers to buy an inventory of your product later.)

After selling your first dozen or hundred or so over a period of at least a month and maybe two and getting some mail-in questionnaires back, you should have a pretty good handle on what the top benefits that interest customers are and what your future sales are likely to be. The retailer will know whether (if your wholesale distributor's price is right, see Chapter 8) the product has a sales/profit level that would justify carrying it in the store (possibly displacing a poorer selling product).

True, you don't have nationwide distribution yet—or even any profits—but you have real information on which to base projections and with which you can convince the distribution channel to handle your product. You still have a lot of work to do if the test retailers say yes to continuing to handle your product—but if they don't, you will be able to wind down with a minimal loss and you will NEVER HAVE HAD TO ASK THE RETAILERS OR DISTRIBUTION CHANNELS TO TAKE A RISK ON YOUR UNPROVEN PRODUCT (a proposition they will almost universally say NO to).

If it doesn't look like the product will sell well enough to justify continuing (i.e., it will not recover your development cost to date plus a good profit at reasonable production runs), but it does sell, you can still sell off any remaining marketing test inventory to recover some of your costs—but don't make any more.

Should I Go On?

How will you or your retail stores know if the sales levels are high enough to justify continuing? Good question and it will vary from store to store but you can do some more math and get a good idea. Often an average of 1 sale a week when you are not there manning a stand and 3-4 sales when you are there for 4 hours will be sufficient to justify continuing. Take, for example, a $6.00 NON-SEASONAL product. Say your manufacturer's sale price is $2.00 for the $6.00 product, with a $1.00 manufacturer's contribution margin and your retailer's projected contribution margin is $2.20 (life's not fair, the retailer gets more than you).

If your previous research has shown that there are fewer than 10,000 retail outlets in the country that might carry your product, assume you'll get about 50% of them 3 years from now. If the research shows over 10,000 outlets then use 33%. Yes, I know that .5 x 9,999 = 4,999 and .33 x 10,001 = 3,300; FUDGE IT—if you keep getting stuck on numerical exactness hire a BUSINESS partner/assistant—don't come whining to me. So let's say there are 10,000 outlets and you project 4,000 store penetration in 3 years.

Seasonal

If your invention is a seasonal product you dramatically reduce your chances of success, stretch the "full" penetration time line to 4 or 5 or 6 years, and your test marketing may be done a year before your first real sales. The focus of this book is NON-SEASONAL products but my advice for a seasonal product is that you try to do focus group tests, with as inexpensive a "product" as possible (perhaps just illustrations), well before your first season rather than doing a real market test with a real product. The narrower your season the worse your odds, i.e., "Christmas" is worse than "summer."

At 4 sales per month times 4,000 we see annual sales of $1.15 million with a manufacturer's contribution of $192,000 and an individual retailer contribution of $105.60 per year. Holy cow, how do retailers stay in business!

Well, imagine now that your product ties up a 6" by 12" shelf space or 72 square inches. The retailer's contribution per square inch is thus $1.47 per year. A typical small retailer might have a 1,500 square foot store with ½ of that devoted to shelf space 6 levels high and would therefore have a total annual contribution of $952,560 at $1.47 per square inch per year. The hell with inventing, I'm going into retail.

Remember, our product is new and unique (and possibly patentable) and we did our initial cost studies well so our product commands a higher than average contribution, otherwise why would the store even carry it? Also don't forget all those other expenses which typically make it hard for a store owner to even keep 5-10% of sales before taxes. Now you see why "traffic" is so important to a store owner but you should also see, for our example, that our "normal" sales level should get us enough money to pay back our startup and market building expenses and leave some left over for profit.

(Just incidentally, the vast majority of retail stores are well below "average" for annual contribution per square inch but the owner doesn't know that till they try to sell the store. This happens because the major chains have larger stores with many more square inches that they put only higher turnover items in.)

Your point in doing in-person sales for a few hours in each store is twofold: one to raise awareness and the second to gather information with your in-store questionnaires. At say 4 sales in 4 hours you only "contribute" $4 to your manufacturing interests but you clearly prove that direct, in-person sales will not pay even minimum wage. This is the usual case.

On the other hand, you may be totally surprised that virtually everyone buys one and in-person sales readily exceed minimum wage. If this is the case, as it was for the original clamp-on, wing twist handle can openers, you may have a product that will be a good door-to-door sales candidate. Such a window will not likely stay open forever but while it does you have an opportunity to use a distribution channel that may reward you much better than the normal manufacturer to distributor to retailer channel.

Try it locally through one or two Boy or Girl Scout or church or other youth organizations before preparing materials for national distribution. A 50/50 sales price split between you and the seller(s) will usually be readily accepted and the door-to-door sales have the added benefit of increasing consumer awareness of your brand/product. If your product is one that will have repeat sales this may be a viable option for a long time to come, but for a one-or-two needed per home type invention the window is likely to be less than 3 years.

Alternatives to Retail Test Marketing

The main problems on this step are 1) your product may be too large or complex or expensive to make a reasonable number of prototypes for actual sale, and 2) if it is not a retail product, having a small stand in a store may not be a reasonable test market approach. These problems can be successfully dealt with as discussed in the next few paragraphs.

I have seen a test market conducted for a sports car where quality, artist done sketches and fairly general design specifications were presented—by a person with credible business experience building race car chassis—to car club enthusiasts. The deal offered was a $3,000 (this was about 25 years ago) prepayment now gets you the car at significantly below (maybe 75% of) what the final price would be set at. The catch was that if things never got to production you stood to lose your $3,000 or whatever part of it couldn't be recovered from corporate asset sales at closure.

There were "sales," development was started, problems were encountered, a second investment of $2,000 was required, a prototype was built, but production was never achieved. The catches were that the near production prototype was significantly different than the original concept and that government regulations required the car be sold as a kit to avoid the costs of meeting crash-worthiness tests among others. In the end all "buyers" were actually given their money back (but no interest) and the 2 partners in the project ate the costs. If a car is the level of your simplest idea I'm not sure where I should have clarified my previous chapters, but the point is, you can do something similar.

You too, can take orders in advance for delivery later but I suggest you get a lawyer involved early and make absolutely clear what the deal is. The preferred, but rare, deal is you give buyers a prepayment discount and hold their money in escrow to be returned in full if production never starts. Your objective is not to get the money to cover your costs, etc., but to find out whether or not a large enough volume of customers are likely to be willing to buy. If it doesn't look like it, stop (and return the money). You'll have to sort out whether your product was worth it at full price, whether the discount customers only wanted "the deal" or whether they would pay full price, whether some thought "Escrow? Ha! I wasn't born yesterday," or whatever.

In other words, you can approach a "Will it sell?" answer but never get it for certain this way.

In the sports car example they just had a sketch but these days you can go with a full video. But be careful; there are dozens of companies vying to

---

 You do sales in advance of production on "special" terms that may not reflect the real market.

---

make 3D renderings of your concept and electronically prepare video tapes of it in action. The <u>pitch</u> is usually that the video is almost a sure fire way to get your idea licensed without you ever having to do the development work; you just pay the service for the 3D rendering and video. Since you have already cut your 100 ideas down to this good one you can make a good business decision on whether to get such a video or not. Evaluate it on its price versus what you need to do to give potential customers a realistic idea of your product.

Trade Shows

For many non-retail products, a display booth at the kind of conference or trade show your buyers attend is the standard way of introducing something. Sometimes here you get caught in the "You have to be a member to join" trap. In other words, you want to have a booth at an industry conference where the industry is defined as all being members of the sponsoring organization—and letting new members in will only interfere with current members' profits (or so the philosophy seems).

Usually you can find an insider friend to help by asking around in the industry. Usually a large percentage of the sponsoring organization members are as ticked off by the exclusion policy as you are (they want to meet people who want to BUY, they don't care who they pay "dues" to). Befriend one and get their help. The problem is you can rarely do this on short notice. If you have to start a year in advance anyhow, you may be able to join for $250 or $500 and have fewer hassles. If you find an insider friend, broach the subject of sharing a booth with them if you want to—they might just be willing to share for a fraction of the booth space fee. Three good places to look for trade shows are Trade Show Central at www.tscentral.com, The Expoguide at www.expoguide.com, and The Job Shop Network at www.jobshop.com.

An alternative is to locate and buy (or copy if it is legal) an appropriate mailing list for the industry. You will need a quality direct mail piece including a brochure and cover letter complete with stationery and envelopes. Send your direct mail piece to several hundred prospects. Follow it up with a phone call in one week and get some feedback on your product, your offer, the prospect's current needs, etc. If you are presuming your product will be a long-term product be careful not to bias your results with substantial special offers and incentives or you won't get a realistic reading of the market.

A Business Focus Group

If it is not a retail product and you don't have production products ready to sell and there are enough potential buyers locally or in a nearby large city, you can announce a focus group with a letter and ask for RSVPs to help you identify the strengths and weaknesses of a proposed new product for the industry. Don't forget to tell them what problem your product solves. If it's an important one to them you'll have no problem filling one or more focus groups of 8-12 people. You get the idea.

In this step you have to get some real buyers and some (positive) "Will they buy?" answers before you commit to full steam ahead. You're still not out of the woods (especially if the question is, for all practical purposes, hypothetical—witness the Edsel and New Coke) but, if you have been honestly accepting and evaluating inputs up to this point, you have cut your risks considerably. Either you have thrown away or substantially modified multiple original ideas or your idea is genuinely proving to be first class.

How Many Mistakes Can You Make?

One reader of the draft of this book asked, "Why don't you put in some examples here of how to change the test marketing so that sufficient sales to warrant continuing are generated?" The answer is, YOU DON'T change your marketing, YOU STOP. It's remotely possible at this point that you have:

1) misidentified the correct target client,
2) not provided the information your market needs to make a buy decision,

3) chosen the wrong retail stores or other channel and/or

4) screwed up your packaging.

What are the odds the idea was wrong versus you failed on those 4 items? The business odds of success after a failure at this point DO NOT usually justify continuing.

That said, of course there is the famous story of 3M's Post-It self adhesive notes. They knew, from in-house use and focus groups, etc., that the product was wanted—but it just wouldn't sell in office supply stores. My guess is their packaging and in-store displays (if any) simply didn't convey the utility of the product. 3M solved the problem by at first giving away packages of the notes. People that tried them re-bought and spread the word.

Last Gasp Marketing—Make it a Promotional Item

If your product is (generally) inexpensive, cute, useful, etc. such that people would like to have one but don't see buying it as an acceptable trade AND it is easy to print logos, text, advertising information on, you might just have a promotional item. A what? They used to always be called advertising specialty items but now are often called promotional items too. They are things you are given for free by businesses just so you will remember, and think favorably, of the giving business. The reason for the different names is to avoid a name conflict with the Advertising Specialties Institute, a for profit company that used to monopolize the industry.

A couple of catches with advertising specialties are that you must be able to deliver fast (2-3 weeks is good but you will also compete with people who can do next day delivery), including imprinting the advertising message, to avoid missing opportunities and your product needs to allow fairly big margins at a low price. Typically the promotional product distributer that routes an order to you receives 50% of the "below wholesale" price the end client pays. Say the client buys 1,000 at 85 cents each. The person that brought you the order would get $425 and you would have to make and imprint 1,000 of them within the other 42.5 cents each in order to make a profit. Far and away the biggest promotional product sellers are under $2.00 each to the end buyer so if your product wholesales by the dozen at $6.00 each item, you will have a tough, but not impossible, go of it.

You will also need to get your item listed with directories of advertising or promotional products to avoid having your marketing expenses kill you. In the great scheme of things the registration/membership fees are inexpensive but they are obviously another expense before (if) you see revenues. For more information on this option check out Bob Merrick's book, the Promotional Products Association International web site at www.ppa.org, the Advertising Specialty Association for Printers (ASAP) at members.spree.com/asap, or the Advertising Specialty Institute (ASI) at www.asicentral.com/noncomp/index.htm. All three web sites have member-only areas that you can't get to and at the latter you'll get a good dose of the attitude that caused competitors to spring up in the industry.

Be aware that the promotional products industry is extremely competitive and not easy to break into as a new company. You may find it much easier to join with a current player in the field or to license your invention to one of the middle sized players that has the right kind of manufacturing for your invention. Do not expect to get rich in a hurry unless you just happen to create the next hot advertising specialty fad item.

If you want a marketing savior don't count on me, but for cash up-front, I'll look at the situation and see—on a one-shot basis—what I can do. As long as you have money to burn there are many "marketers" who will "help" you (cash up-front please) till you're sure the product is dead (or have convinced yourself that everyone IS out to get you).

20. Your product only sold when there was a person doing the selling and sales were not high enough to very easily cover the labor.

21. None of the stores that initially offered your product during the test marketing are willing to continue to carry it...

UNTIL you fix some specific, easy flaw (such as using anti-pilfering packaging).

22. The projected contribution margins for you, distributors, and/or retailers at the proven sales volumes at the test price(s) will in no case be sufficient to warrant the handling of your product over other products.

Explanations and scoring suggestions:

20. <u>Your product only sold when there was a person doing the selling and sales were not high enough to very easily cover the labor.</u> In-person selling almost always gets higher results than just shelf selling but it usually only is worthwhile on higher priced items where the knowledge of a salesperson is a benefit to the buyer. Quality stereo (or multi-channel surround sound these days) exemplifies such a situation.

 0%—Sales were fantastic in-person and products continued to fly off the shelf when no one was there.

 2%—Sales were good in-person and ran at a higher per square inch contribution than the average item in the store.

 5%—Sales were good in-person and the per square inch contribution ran just a bit lower than the average item in the store.

 60%—Sales were passable in-person when benefits could be explained but the contribution per square inch was close to the worst in the store.

 85%—Sales were poor in-person and only a few sold off the shelf despite window and in-store promotional signs.

 99%—Sales were terrible in-person and none sold off the shelf.

21. <u>None of the stores that initially offered your product during the test marketing are willing to continue to carry it.</u> ASK FOR the specifics of why they won't carry it. Be aware that "It's not the type of product for our type of clientele" is probably a euphemism for "IT STINKS."

 0%—All test market stores demand quantities of the product beyond what you can immediately supply.

 2%—All stores tell you to come back to them when you are in production.

 10%—Most stores want to continue the product regardless but tell you a lower price point would be much better.

25%—All stores will continue to carry the product but only if you will fix some easy flaw (such as using anti-pilfering packaging).

 50%—Most stores won't carry the product until a couple of significant problems are addressed.

 75%—No store will carry the product until significant problems are addressed.

 100%—No store will carry the product for reasons that are essentially unsolvable.

22. <u>The projected contribution margins for you, distributors, and/or retailers at the proven sales volumes at the test price(s) will in no case be sufficient to warrant the handling of your product over other products.</u> EVERYBODY in the distribution chain has to make WHAT THEY CONSIDER a decent (or better) profit from the product or the product won't be moved through the chain at all.

 0%—Everybody in the chain comes out with a better contribution from your product than they get from their average product.

 5%—Everybody in the chain comes out with about the same contribution they get from their average product but you should be able to improve that for them when production volumes get ramped up.

 30%—One point in the distribution chain gets squeezed a little but the others come out okay.

 75%—This product's contribution for almost everybody in the distribution chain does no better than the worst 25% of their current products.

 99%—One or more points in the distribution chain do not receive a contribution that warrants their handling of the product.

STEP 6—Create the Marketing Materials and GO!

Real U.S. Patent—and PCT?

(Be forewarned, this gets really dense.)

Patent considerations for this step: Well before your 1-year Provisional Patent Application clock, if you filed one, and (hopefully) 2-year Disclosure Document clock expire, have your patent attorney get started on the U.S. Patent Application and/or a PCT (Patent Cooperation Treaty) Patent Application. Whoa! PCT? Isn't that...international?

Yes it is, and your multi-country patent protection, if it's economically viable in the first place, may be much less expensive in the long run if you start by filing your PCT application with the U.S. Patent Office along with the U.S. Application. That's right, the same office that takes your U.S. Patent Application accepts and processes your international patent application initially. Eventually, yes, you will have to pay fees to foreign governments and probably attorneys. But you will be well down the profit path by then or will be able to abandon your application knowing that your product really doesn't sell and any patent would likely be commercially worthless.

Before deciding U.S. or PCT or both, do some real market research on the potential of foreign markets to be sure you understand where it is worthwhile to sell your product. For some clues on how to start this see the Foreign Economic Market Information in Appendix G.

Also get more information about the PCT application process from the World Intellectual Property Office (WIPO) at www.wipo.int. WIPO Publication No. 433(E) is available at www.wipo.int/eng/main.htm (click "PCT system" then "Basic Facts about the PCT") and is a good starting point even though it is an extremely abstruse document (Flesch-Kincaid Grade Level 16, Sentence Complexity 80, Vocabulary Complexity 63). The *PCT Applicant's Guide* is also there and can be looked at online or printed or it can be ordered from the USPTO. The top 8 economic countries, as well as 95 others, are all PCT members (the complete list is on the WIPO site).

After you are educated in PCT filing ask your attorney what the pros and cons of just starting with the PCT filing are and what they will charge for <u>preparing</u> the PCT Patent Application. If they quote up to about 10% more than for a U.S. application, they are probably honest. If they quote up to 25% more, make them justify it with SPECIFIC reasons relevant to your invention. If they quote more than that my first inclination is to just have you shoot them on the spot but you'll probably be better off if you just get up and leave—then maybe announce their quote at your next local inventors' club meeting.

Since you have a solid product in hand AND WON'T BE MAKING CHANGES while the patent is being written, you may also want to ask your attorney to give you a fixed price on at least the writing and filing of the patent application. If they have done a patent search and feel the "prosecution" phase of the patent process should be fairly predictable, they might even quote you a fixed price on it too. If they offer it for a bigger fee than you think is reasonable you might want to insist that phase be done on an hourly basis. If they offer a fixed price on prosecution they will likely exclude interference costs. Decisions, decisions!

Prosecution? Interference?

The "prosecution" phase of the patent process can be thought of as the time from filing to award or "final action," but the effort and expense don't typically start till the first "office action" by the patent examiner which will probably occur 6 months to a year after filing. An "office action" is just a letter stating the USPTO's action or current decision and the first office action is typically rejection, with the reasons stated, of one or more claims in your application and requests for technical corrections to the submitted application. EXPECT THE FIRST OFFICE ACTION TO BE REJECTION.

It is usually not that you or your attorney failed or goofed up the application. It's just that another set of eyes is looking at the application from different angles. When your or your attorney's adjustments are made or successful rationales to overcome objections are forwarded you will likely end up with a stronger patent than was initially filed.

"Interferences" occur only in about 1% of all patent filings and it is these situation that make your records very important. Interferences are much more likely to occur in active, leading-edge technology areas. An interference occurs because the patent examiner finds another current application, or an issued patent that was issued within one year BEFORE your priority date (some USPTO documents say "filing date" but are probably just copying pre-1995 terminology), that claims substantially the same invention as your application. Each party must then present evidence of the facts regarding conception and reduction to practice and the USPTO's Board of Patent Appeals and Interferences will make a decision for one party and against the other.

But it ain't necessarily over yet. The loser can appeal to the Court of Appeals for the Federal Circuit or file a civil action in the appropriate United States district court. In big company interference situations where the USPTO decision is reasonably contestable, deals are often worked out for cross-licensing of the contested and other patents rather than engaging in a court battle. Think about that option regardless of which side of an interference decision you are on. A creative partnership of two parties working together is likely to be able to produce far more that two parties fighting "to the death."

When your patent practitioner is done preparing your patent and is ready for you to sign it DON'T. Take at least a couple of hours and preferably a day or two and read it closely looking for spelling, grammar, capitalization, punctuation, misidentified references, missing reference numbers on the drawings, etc. It is your responsibility to be sure it is correct. DO NOT TRY TO REWRITE IT YOUR WAY. If you have questions, get explanations. If your patent practitioner is competent and conscientious this last check will be tedious but easy and painless. If you find more than 3-5 nits or 1 or more substantive errors make sure they are corrected then consider a different firm for your next patent. Problems with nits likely won't cause a judge or jury to decide against you but they may be chinks in your armor that someone might try to exploit.

Issues you might be concerned about in making your U.S. vs PCT choice include what your "priority date" is, extra fees, when publication occurs, and export licensing. If you file a U.S. Patent Application and/or

PCT Application and had filed a provisional application, you have your choice of establishing the priority date as either the date the provisional application was filed or the actual date of your U.S. Patent application. If:

1) foreign patents won't be beneficial to you

2) and you have date-of-conception proof outside the provisional application

3) and NO publication more than a year ago

4) and your Provisional included information you will not be including in your full patent application,

5) or omitted key information that will be,

then the actual patent filing date is probably best because your U.S. patent, if granted, will not cause disclosure of the extra material in your provisional application (if your product discloses patentable material that isn't in your real application and perhaps was not in your provisional, get good legal help ASAP).

If you had public disclosure (such as test marketing) after your provisional application and you want foreign patents then you need to use the provisional filing date. If you file a PCT application, the fees due the Patent Office are more than 4 times what they would be for just a U.S. application (PCT ≈ $1,557 plus $105 for each place you plan to file in [max $1,050 for 10 or more places]; U.S. ≈ $345; PCT & U.S. ≈ $1,652 plus $105 country charges up to 10) but your long term costs should be lower.

The initial PCT fees (which are subject to change WITHOUT being posted on the Internet since they fluctuate with currencies) get you through the Search Phase only of the International stage. Examination, National Stage, and other fees will be required later. For current PCT fees call the PCT Help Desk at (703) 305-3257.

Filing of either a U.S. or PCT application also automatically results in a request for a license to file for foreign patents if the invention was made in the U.S. You can be denied the right to file for foreign patents. Notification of the granting or denial of your license will be returned with your notification of receipt.

146 How to Determine If Your Invention Is Profitably Marketable

If you go with only a U.S. patent application, it is only published when the patent is granted. If you go with a PCT application, your invention will be published in 18 months if the application, or foreign applications, have not been withdrawn. In other words, it will probably be public before you get any patent. I hear a lot of inventors rail against that publication but my feeling is "So what?" If you are following the steps in this book, your product should be on the market and therefore public (and making profits) long before that publication. If, **with USPTO permission, you file directly with some foreign governments** your invention may be published in as little as 6 months.

Another advantage to PCT filing is that in 5-6 months a preliminary international search for relevant prior art patents will be done and the patent office will return to you a report indicating what was found and whether they believe that patent denial will occur due to the found prior art. Armed with such information you will be able to make decisions on if and where to proceed.

If you didn't do a Provisional Patent Application, your 1 year clock started with your test marketing or other significant non-confidential disclosure and you may no longer be able to get foreign patents. The Patent Application (or a stalling provisional) must be completed and filed before the deadline otherwise the patent, should you get one, may be later declared invalid.

Be aware that in 1998 there were 243,062 U.S. patent applications filed and only 147,521 patents granted. Since patents take a while to process a better year to look at might be 1995: 129,749 out of 228,238 patent applications filed in 1995 have been granted patents. That's 56.5%. Long-term statistics show that only about 9 more percent of the 1995 filings will be granted as patents. If you've done your work carefully to this point the odds are you'll be among those being granted a patent.

If you filed a Provisional Patent Application and are going for foreign patents you must get going on them at this time, preferably via the PCT Application. If you didn't do a Provisional Patent Application **and** all previous disclosure was "confidential" you can probably safely hold off on foreign applications for a few months after you file your U.S. Patent Application. The maximum may be 3 or 9 months depending on the country but for all PCT countries it's 9 (that gives your patent practitioners

James E. White   STEP 6–Go!

3 months to get everything done and submitted). **You should NOT wait for a U.S. Patent to be granted.**

If cash flow is a problem you should work out with your attorney a plan where all appropriate deadlines are met but minimal expenses are incurred. One trick, for example, is to file informal drawings with your application rather than go to the full expense of formal drawings. While I don't recommend the practice it can save some early cash (and even later cash if the Examiner requires new or revised drawings). Another catch is your informal drawings will need to match what you expect your final drawings to be. The good news is that if your patent is denied you will never have to foot the expense of the formal drawings.

With a product that is proven to sell you may even be able to work out a payment plan with your attorney so that they don't need all the money up front. For foreign patents you will usually get foreign law firms involved and letters of credit and payment guarantees may get tricky, so do some advance homework with your attorney.

Keep in mind that a patent in a country can be used to exclude (at your trouble and expense) <u>legal</u> knockoffs made in a country in which you do not get a patent. Also be aware that some countries' patent laws require that you manufacture and/or license the manufacture of your invention in that country to keep the patent in force (the grace period is typically 3 years). Finally, be aware that many governments (including the U.S.) require additional payments over the life of the patent to keep it in force. The payments are trivial for a smashing success but can be a burden for a marginal product. You, of course, have avoided it, but those payments are a real killer for inventors who file the patent application first then limp along in development or have to try to create a market or find a licensee for their unproven invention.

Do You Infringe?

In this step you may also want to have a check done to see if there is anything your patent might infringe on. In examining applications for patent, no determination is made as to whether the invention sought to be patented infringes any current patent, only that the patent office believes the <u>claims allowed for your patent</u> are new. An improvement invention may have patentable claims, but it might infringe a prior unexpired patent for the invention improved upon, if there is one.

There is no requirement that you check for infringement so, as a business decision, you may believe the risks of infringement are low enough and that you will be able to amicably negotiate with anyone who claims infringement after your product hits the market. You may want to hold off on ferreting out potential infringement until after your patent is issued, then have the study done and preemptively make an offer to the people you possibly infringe on. Follow your gut instincts on what is "right" and you will probably be okay but there are no guarantees. It's the old "no path is without peril" quandary. Your own patent search should have helped you avoid the potential for problems in the first place.

A legal legal maneuver you can use even if you don't believe your product will be granted a patent is to apply for one anyway. Only about 65% of all applications mature to patents anyhow so you don't need to think of yourself as a sneaking loner way out on a limb. Your application still must be for your invention or you will be fraudulently signing the oath when you file. (If you "accidently" find out what the penalty is for this or how frequently it is applied, let me know.)

With the filed application you can use "Patent Pending" until you give up the fight. The advantage of the trick is that it MAY (no guarantee) delay entry of knockoff or competing products. If you are extremely lucky you may even get a patent that is never challenged or invalidated. The disadvantage, of course, is it costs you time and money for the filing, etc.

Again, do not forget to include "Patent Pending" on your product and/or materials if one is. And don't forget to switch it to "Manufactured under U. S. Patent No. 999999999" and/or whatever is appropriate for foreign countries when your patent(s) are granted and where you are distributing. If you are only distributing your product in the United States "Patent 9999999" (using your patent number of course) is the simplest form of notice required if you wish to be able to collect damages from infringers.

So now you know, PATENTING is about the last thing you want to do with an invention idea. It is, of course, about the first thing you want to do with a proven PROFITABLE new product. But, if you've followed the steps, you've got about 9 months of profits to receive before incurring the major patenting expenses.

CAUTION: After your patent is granted "the hawks" will descend on you with mailings of congratulations and offers of "opportunities." These hawks know that the vast majority of individual inventors believe that "success" is assured with a patent. They don't know that you will already have proven to be a financial success and that the patent is just "icing" to you. They will offer you fancy plaques, marketing assistance, presentation of your patent to "inside" contacts, *ad nauseam*. Make no mistake, they are not sincerely congratulating you, they are targeting the suckers who probably bought $5-10,000 worth of worthless patent. They don't make their money from success with your invention but from your up-front payment to them.

WARNING: A particularly pernicious form of this "hawk attack" may come 3 or 6 months or even a year after your patent grants. You will receive a notice that your patent is being referenced or cited (or something similar) in a (or their) new patent. For immediate info all you have to do is send them $50 or $100 or whatever. IGNORE THEM COMPLETELY. They want to raise the fear in your mind that your patent will somehow be damaged and the only way to protect yourself is to pay them. Your patent likely cites prior art patents—and your patent didn't damage those previous patents in any way, did it?

Just wait quietly and see if their patent ever issues, it is unlikely that such a thing will happen. On the other hand you might want to watch specifically for patents where manufacture of an "improvement" would include infringing on your patent. You don't have to constantly do patent searches or subscribe to the *Official Gazette*, for typically less than $3,000 a year you can have a service keep their eyes open for you. Your patent practitioner will very likely know 2 or 3 services they trust, you don't have to accept one based on just a mailing.

You can also buy "patent insurance" for enforcement and/or defense of your patent which will usually pay litigation expenses up to some cap should your patent ever be infringed or challenged. This insurance can be very expensive or quite reasonable but you will have to negotiate your own policy and its fees. Again, your patent practitioner will probably know trusted services. For a worthwhile patent that is generating $100,000 or more a year in profits these expenses can be small and worthwhile.

The guts of STEP 6 will be for you, or someone working for you, to complete all your marketing paraphernalia from packaging to consumer display ads as appropriate. It will likely be stretched out over a number of your early production runs as you build up distribution and sales. Before you go full scale on production you will need to at least get a UPC product ID (current fee $500) (www.uc-council.org) and complete your packaging. You will also need to prepare whatever materials are expected by the distribution channel (catalog sheets, packing boxes, quantity discount sheets, etc.).

NOTE: your UPC application fee will actually buy you an ID and a block of numbers. Your ID will always be part of your UPC code but you can assign the numbers to your products as you wish as long as you do not use the same number for 2 different products. An individual widget must be assigned a different number than a 6-pack of the same widget. Since accurate reading of your UPCs is critical, you should use a service specializing in them to produce the originals for your product label/package printer.

Copyright

Copyright your commercials, advertising, package printing, instruction sheets, manuals, etc. Technically each of the "works" mentioned in the previous sentence is copyrighted the instant it is "fixed in a tangible form" and you own the copyright—even if you don't provide notice.

If you retain an individual or ad agency to create materials be certain your contract or agreement (in writing) spells out that the materials created for you are "works for hire" and that you retain the copyright.

To publicly declare your copyright (i.e., "give notice") all you need to do is place the word "Copyright" or the "©" symbol (or both)—"(C)" is NOT valid—followed by the year of first publication and a generally recognizable form of your name or DBA or corporate name. It also won't hurt, but it is not required, that you include "All rights reserved." (See the back of the title page of this book for an example of copyright notice.) The "All Rights Reserved" used to be important (but probably isn't any more) for full foreign coverage and where alternate media might be used such as a movie from a novel.

Revised versions of your works should have the correct new year of publication in their copyright notice, not (just) the year of the unrevised original version.

To "perfect" your copyright on a specific item you must register your copyright with the Library of Congress Copyright Office (www.loc.gov or specifically lcweb.loc.gov/copyright). Each copyright registration with the Library of Congress requires completion of a form, a check for $30 (or so depending on type), and usually one or 2 copies of the material being copyrighted. It may take you a while to dig out the information you need but it is all there at the Internet site.

A U.S. Copyright is recognized by over 190 countries including most that are likely to provide viable economic markets for you so you only have to register your copyright and pay the fee once for each item. For an individual the copyright is good from the moment the work is in "tangible form" through 70 years after the death of the individual. For a "work made for hire" or with no identified human author either in the work or its registration, usually for a company, the copyright is good for the lesser of 95 years from first publication or 120 years from creation.

Failure to provide a copyright notice with publication, especially if it can be shown that the failure was not accidental or that, even if it was accidental, no attempt was made to correct it, can completely invalidate your copyright so be very careful. If your copyright is not filed with the Library of Congress the United States will refuse to become involved in any disputes arising out of your copyright claim. If your copyright is filed with the Library of Congress the United States will lend its support to your case (but it won't make it for you) and you may be eligible for treble damages should you prevail. Customs intervention may also be possible. In my opinion, the $30 (or so) fee is well worth it.

Relative to your product, the most valuable place for copyrighting is probably on instructions that are provided with the product. It is far more likely for a knockoff artist to copy those (or to try to get away with a "derivative" version of them) than to copy your packaging unless they are fraudulently attempting to sell their product as yours.

Liability

You also need to get your liability insurance lined up. In fact it might have been a good idea for you to have talked liability and liability insurance over with a good business insurance agent before you did your test marketing. Essentially that is a judgement call on your part and will be heavily dependent on the risk of injury or other calamity that your product might incur. Insurers can also be very helpful in specifying what kinds of warning labels you might need on your product and what warnings and disclaimers you might need in your instructions and/or warranty.

Be careful and don't overdo it though. Make your product as safe as practical, i.e., would it pass a "reasonable man" test? (sorry ladies, I didn't invent that phrase). Build in good safety in the first place then add labels and instructions that are genuinely in the best interests of your customers. "Warning: do not fall off ladder" may be amusing the first time you see it, but it won't save your butt if they fall off because the ladder breaks.

The fun of inventing is over, now you have to get into the hard work of making a profit from it. If you have followed these steps you should have a far better chance at that than most inventors. After all, your product sells—and will sell profitably. You already know that!

For more on what you need to create for your marketing materials see Chapter 7 "Before You Enter the Marketplace." That chapter also concludes with a list of all the "marketing" and "invention marketing" books mentioned elsewhere in this book.

 23. You just can't stand the fun of building a company and collecting your first profits.

Explanations and scoring suggestions:

Score this a 0% (☻) please, otherwise why have you been working so hard?

Actually, if you really don't want the hassles of a company and doing Step 6 yourself, you might want to immediately start looking for licensees instead of proceeding with further production and marketing. You'll be in the driver's seat but you won't really have your seat belt cinched down till you've proven the test market wasn't a fluke due to some accidental good luck.

You might also, if you are the technical type that loves inventing but doesn't tolerate business very well, be able to recruit an entrepreneurial type person onto your team. The entrepreneur will delight in the challenge of gathering profits from your invention and will leave you (mostly) free to work on your next invention. You will, however, have to give up most bragging rights to the success of your invention (and a major chunk of profits) because the entrepreneur will likely (and erroneously) primarily see the profits as a result of their efforts, not the results of your invention. Actually, both are necessary—but I think, and you'll probably agree, the invention is the most important contributor to success. One or two unique brains cook up the invention idea but virtually any entrepreneur can successfully sell a product that <u>correctly meets a need at a profitable price</u>.

"STOP IF" Scoring

Add up all of your percentage scores. If the total is <u>greater than 60%</u>, set that invention idea aside and tag it with a DO NOT RESUSCITATE order. If the total is <u>40 to 60%</u>, set it aside and let your subconscious work on it a while. If the total is <u>20 to 39%</u>, get a second opinion. If it is <u>5 to 19%</u>, proceed with caution concentrating on the higher probability "STOP IF" questions first. If it is <u>under 5%</u> proceed merrily on your way—but don't forget to re-ask the "STOP IF" questions when new information makes that appropriate. During development, for example, it is not unusual to discover that manufacturing is either impossible or too expensive relative to what customers might be willing to pay.

Keep in mind that you should NEVER get to a "final" score at the end of STEP 6 if you contentiously kept a running total score at the end of each preceding step and came up with any running score over 60%. You should have already stopped. If you proceed like most self evaluation systems encourage you to do you will often fall into the trap of working hard to improve a particular score that doesn't amount to beans in the long run. For example, the "If I can just get a patent." mentality kicks in.

Patenting the 7,385[th] toilet seat lifter might, in your own mind, give you a "perfect" score (0% in my system) but the reality is the easy bypassing of your patent should still result in a patent <u>protection</u> chance of failure score of at least 50%, not 0%. The guy that "invented" belt mounted "holsters" for carriable sized cellular telephones did not (and very likely couldn't) get a patent yet still had a successful 3 year run before competitors "copied" (not stole) the idea and flooded the market with cheaper (and often inferior) knockoffs. His BUSINESS decision

---

> was not to score the Patent question. It was a good, market-driven, decision.

STEP 7—Keep an Eye on the Competition

This isn't really a get-the-product-to-market step; I just kind of added it as an afterthought. If you licensed your invention you need to be extra keen and careful on this one. Every few weeks you should go shopping in your favorite stores <u>and in less familiar stores</u> for your solution to the problem <u>and its competitors</u>. When you travel anywhere for any reason also make it a point to go shopping. If you have a patent and find a knockoff product that you think violates your patent and/or infringes your trademark, buy it. Also ask for (and write down if it's not on your receipt, etc.) the name and address of the store you bought the product in. DO NOT get mad and rant and rave at the store clerks or the management—in all probability they are innocently passing on a product introduced elsewhere in the distribution channel. DO take the product to your Patent and Trademark attorney (or your licensee if they are responsible for pursuing infringers).

If your attorney or your licensee's attorney do not believe there is an infringement you may want to start thinking up ways you can out-compete your competition before they wind up with a very lucrative toehold in the market that they won't want to give up. If your attorney does believe there is an infringement, get their cold hard rational assistance in tracking down both the source of the infringing product and all current distributors and retailers. Your attorney will want to send a strongly worded cease and desist letter but you might be better off going with a polite, informative letter to all but the infringing manufacturer. Regardless, you will need to follow-up and take further action if warranted. Amazingly enough, simple letters are usually all that it takes to defend your rights (at least in the U.S.).

CHAPTER 6
Pitfalls to Avoid

1. Free invention evaluation services. The kind you hear about in radio commercials or in display or classified ads in magazines with a technical bent. These offers virtually always respond that your invention has potential and that their staff and expertise can be put to use to "present your idea to industry"—for a fee, of course. I would recommend that you team up with a friend in another town <u>if you want to use one of these free services</u>. Between the two of you come up with two inventions and write them up and submit them separately (one for each of you).

Compare the results you get back. Are they almost identical—right down to the amount you have to pay to proceed? Do they seem to ignore your submitted inventions in favor of generic, positive but waffling (or even ambiguous) statements? Do they guarantee something that is easy to do (submit your idea to 100 manufacturers for example) for a fee substantially higher than what you could do it for yourself? If "submit" is undefined in their response you can bet it is the simplest possible, perhaps a cover letter with your idea attached and "submitted" for little more than the cost of first class postage.

They also often say they prepare a "brochure" or "portfolio" for you but most of the time these are simple, 1-page flyers with a big illustration and a few (often exaggerated or erroneous) bullet points below the illustration. Ask to see a few. If you are told they are "confidential" and you can't see them, run. If they give you one or two flyers and tell you the inventor opted for the least expensive "brochure," run. In other words, if "buts," and "ifs," and other "hedge" words are their predominant form of answer, RUN.

If it costs you money with minimal guaranteed effort on their part then you are probably dealing with a firm whose real business is stroking the egos of inventors, not "presenting your idea to industry." If you need your ego stroked because it got crushed in the "Steps for Idea Development" section you may need to reevaluate the MARKETABILITY of your product. Assuming you feel it is still <u>profitably</u> marketable, you had best consider whether the $5-15,000 you would spend getting your ego stroked might be more profitably spent on completing development and arranging your own manufacturing.

Also be very wary of "developers" who will complete development for fees (that often more than cover their costs) plus a percentage of future profits. The "development" will likely result in exactly your initially described invention which won't work—till you provide more money and the ideas to fix it, etc. Once you've paid for all the work to get it working, which you would have had to do anyway, they still get a piece of the action if YOU can profitably sell it. Is this fun or what?

2. Starting with a patent attorney. While you may get one that will be helpful it will be because they are a nice person, not because they are a patent attorney. If you show up in a patent attorney's office with nothing but an idea in your head and they don't bust out laughing it's because they have very good self-restraint (much more than me). A patent CANNOT be gotten on an IDEA, only on "any new and useful process, machine, manufacture, or composition of matter, or any new and useful improvement thereof."

The patent attorney's job <u>is not</u> to help you with your invention, adjust the design so that it works, suggest alternatives, etc.—their job is to DESCRIBE in broad but unambiguous terms the EMBODIMENT of <u>your</u> invention (whether it yet exists in physical form or not). If you start development while they are writing and discover changes that are necessary for your invention to work or that otherwise improve it, you will pay for the repeated rewrites (and repeated drawings if you are paying to have them done).

My advice to you, and I believe most experienced inventors would second this, is to work out the simplest design you can <u>that works</u> **<u>and that can be manufactured</u>** and only then approach your patent attorney. If you really must start with a patent attorney, I suggest you pay them for an hour's worth of time to help you prepare and send your Disclosure Documents to the U.S. Patent and Trademark Office with your $10.00 fee and the filled out "Disclosure Document Deposit Request" form (or equivalent cover letter).

That will give you 2 years in which to proceed with the knowledge that you have some *evidence* for the inception date of your invention. You still DO NOT have patent pending status and your IDEA is still not protected. All that has possibly been established is a date for "first to invent" assuming your Disclosure Document invention is fairly similar to your final invention. Note also that the Disclosure Document is meaningless unless you <u>pursue</u>

your invention **and file a patent application**. You cannot wait for something similar to turn up on the market then hold out your hand for money.

Bob Merrick suggests that you start with a Provisional Application for Patent because it is still inexpensive ($75 filing fee) and because it does give you "Patent Pending" status and protects your foreign patenting rights. You must file the actual Application for Patent within one year or the provisional application becomes abandoned by law.

The problem here is not the $75 expense but that if your final Application for Patent deviates too significantly from the provisional one, the patent office can determine that your actual application does not "have support in the provisional application" and therefore does NOT get the priority date of the provisional application (and thus preserve foreign filing rights). If you published info on your invention or began manufacturing or offering it for sale before filing the actual application you may lose some (particularly foreign) rights. Talk over (for appropriate fees) your plans and changes with your patent attorney.

My opinion is that you should initially start by going the Disclosure Document route. Anytime your invention changes substantially during development spend the $10 on another Disclosure Document. Once you've seen one done correctly you should be able to do your own. Do not publish your invention and do not offer it for sale or start (non-prototype) manufacturing till after you have filed at least a Provisional Patent Application on your **DEVELOPED** invention. And, as always, diligently pursue the development of your invention.

The Provisional Patent Application does not need to be filed until just prior to offering your fully developed product for sale (or "publishing" it). That will give you a year's breathing room before the actual application is required to protect your rights. (You can also allow the provisional application to be abandoned if the product isn't profitable and never suffer the expense of the full application.)

Attorneys will likely always, correctly, caution you that filing for the patent NOW is the safest route. That is certainly true from the standpoint of irrefutable proof for an early date of invention. It is also far and away the best option for the attorney's pocketbook. If you change your invention substantially during development, they get to do "Continuation-in-part" applications for additional fees. Using the assumptions that the average attorney's fees are $6,000 per patent, and 49 patents out of 50 do not make any money, the average patent attorney will collect $300,000 for each commercially successful patent. That ain't too bad a haul.

3. Expecting to just license it and collect royalties. To get a manufacturer to license your invention will often require that you provide some convincing evidence that it will be profitable for them to do so. Your projections on paper are probably not going to be very convincing without some proof of sales and proof of manufacturing and distribution costs. If you have no proof you may still be able to squeak by if you can show convincing numbers of the size and viability of the market for your invention, including identification of your competitors and the pros and cons of your invention relative to their products. This is best done from the perspective of YOUR BUYERS; your inventor's perspective is irrelevant to product success.

You may need actual surveys of buyers of competitors' products that clearly identify the competitor's strengths and weaknesses in your buyer's eyes. Your invention must exploit the weaknesses without falling down on their strengths in the buyer's eyes. How to go about preparing a convincing report for prospective licensing manufacturers is a book in itself. YOUR results from each of the steps you've completed by following the process described in this book will provide a very good start.

Your invention will likely also have to have some protected uniqueness that is IMPORTANT TO THE CUSTOMER and will make it viable for your licensee to recover their costs, and to profit, after paying your royalty, without undue risks of a competitor providing a similar product without having your royalty costs. The best example I know of here was written up in the December 1996 issue of Inc. Magazine (use www.inc.com for access to many excellent business articles).

An inventor had an IDEA that a giant paperclip would sell. (They had this IDEA after someone requested that their company, a spring manufacturer, make one as a special order.) Anyway they "invented" the giant paperclip (about 4 inches long and capable of holding about 100 sheets of paper while staying "flat") and were awarded a patent on essentially the two new claims for their "invention": 1) the clip's arms (the straight ends) extend the length of the clip and 2) the clip (they call it SuperClip) is made of a high-carbon spring steel. (Note: Why the U.S. Patent Office did not consider this "new" material for a paperclip "obvious" is anybody's guess. My guess is the patent examiner figured the claim was a non-(commercial)issue and, in order to meet the quota of completed patents, gave up fighting the claimant.)

James E. White

The first year sales exploded to $400,000 as Office Depot and other large chains placed orders. The second year sales dwindled to $200,000 as the knockoffs hit the market. The inventor's own market research showed that consumers believed a fair price was about 25 cents a clip yet they had to sell theirs at about 50 cents a clip to make a profit. The knockoffs all came from Taiwan and could be sold for 99 cents per pack of 5. Consumers bought (and are still buying) hundreds of thousands of the inferior (according to the inventor) knockoffs because the "better" features (the long arms and high-carbon steel) were (and still are) IRRELEVANT to the buyers.

Incidentally, the Wal-Mart people—the ones that bring you the WIN program and the commitment to buy American ("whenever pricing and quality are comparable to goods made off-shore")—sell a knockoff product. The inventor has spent about $150,000 in patent attorney fees. (Somehow I cannot imagine an honest and ethical patent attorney accepting that much money unless the inventor beat him or her to a bloody pulp and insisted they take the money and do whatever. Also, see Pitfall 8, "Hiring Hungry Help" for more on this.)

"Well, What if I'm smarter than the average bear and I do get a license agreement?" Well then, prepare to keep your day job forever—but plan on a nice retirement too! Let's say that you did get a license agreement and that your product takes off. After it ramps up, consumers buy at a steady rate of about $3,000,000 a year. How much of that goes in your pocket as a royalty? Let's see, if you get a royalty of 3% the answer is $20,000 per year.

Your math is a bit rusty but you don't get an answer that agrees with mine? Your 3% royalty will typically be on the MANUFACTURER's sale price, not the RETAIL sale price. The manufacturer's sale price will be about one third of the retail sale price or $1,000,000. That gives you $30,000 of which your IRS and state treasurer take at least a third leaving you with no more than $20,000 for your successful idea.

Incidentally, plan on that nice retirement only if you successfully invest your $20,000 per year rather than spend it for each of the 3 or 4 years your product is successful in the marketplace. Sure, some products go the life of the patent and beyond, but in general a product is obsolete in less than 5 years. Do your own math and be realistic about it.

The typical range for royalty percents is 3 to 7 percent going as low as ½ percent and as high as 15 percent. Of course there are exceptions, but what the royalty actually works out to is typically 10 to 20 percent of what the manufacturer's profit would be if they didn't have to pay the royalty. Remember, that is profit, not contribution margin. Also be aware that your royalty percent will be <u>higher</u> the further you've gotten in the development and marketing process. If you've gotten through Step 5-Sell a Few, and proven profitable sales, you'll be in the driver's seat when negotiating a license.

If you've only had the idea, you'll likely be locked in the trunk. The best discussion I've seen of invention licensing in books for inventors is Thomas Mosley's in *Marketing Your Invention*. The bottom line of his advice is: write win-win license agreements. If you don't, you'll likely never get a licensee.

4. Forgetting to do expected return calculations. When I went through the College of Commerce and Business Administration and later the MBA program at the University of Illinois I learned the basic rule of thirds relative to product price. The rule of thirds is that the manufacturer, distributor (or distribution channel), and retailer each contribute about 1/3 to the price of a product that you see on the retailer's shelves.

In other words the retailer must mark up his purchases by about a 33% margin in order to receive enough to cover overhead expenses (labor, facility, etc.) and still leave (hopefully) about a 10% profit on which to survive. The manufacturer typically must mark up their direct manufacturing costs by 66% in order to cover overhead (indirect labor, factory, capital costs, etc.) and leave a profit. The distributor typically expects 33% of their added third of the retail price to recover just shipping costs.

Both the distributor and the retailer have marketing expenses they also must recover in selling the product. Too often an inventor will only look at direct manufacturing costs and figure they should get the bulk of the rest of the retail price, excepting a little profit for the distributor and retailer. Ha! Do the numbers all the way to the consumer but figure your profit based on your manufacturing cost plus your <u>other expenses</u> (Yes, you will have them.) spread over the expected volume.

The "rule of thirds" numbers quoted above are, of course, wrong for the particular industry your invention fits in. If you must, at this stage, have better numbers go to the Robert Morris & Associates web site (<u>www.rmahq.org)</u> and order their *Annual Statement Studies Book*, only $129 for non-members. They list over 500 SIC codes you can get average numbers for.

Unfortunately most of their numbers are based on balance sheet data rather than expense data, but things like Operating Expense percent and All Other

Expenses percent are there as well as interesting numbers such as Gross Profit, Operating Profit, and Profit Before Taxes. One of the things you really want is "commissions" which is a Schedule C or form 1120 deduction line item but is rolled into "Net Sales" in this and other numbers sources and is usually not broken out separately in annual reports. If you wish, for more money you can get regional breakdowns also. You can also get, for $59.95, right there online all the data for 1 SIC code. Their publication may also be available at larger libraries or Small Business Administration offices. Or, it might be time for you to have a little chat with a prospective accountant or banker for your future firm. These offices often have the Robert Morris & Associates materials or similar ones such as the following:

> *Industry Norms and Key Business Ratios: One Year, Retailing,* Dun & Bradstreet, www.dnb.com (Net Profit After Tax is included);

> *101 Business Ratios: A Manager's Handbook of Definitions, Equations, and Computer Algorithms,* Gates, S., 1993 (this book doesn't have numbers, just how to compute ratios and some pithy assessments of why they are useful);

> *Almanac of Business and Industrial Financial Ratios,* Troy, L. (good breakouts by Cost of Operations, Salaries & Wages, Interest, Taxes, Depreciation, and Pensions);

> *Advertising Ratios and Budgets,* Schonfeld & Associates, Inc. (multiply the ongoing firms' numbers by 5 for your startup number)

To get all the right numbers look up the appropriate manufacturer, distribution, and retail categories. In most cases these days you'll find that significant advances in warehouse and distribution automation have reduced the distribution channel to less than ⅓ of the final retail price.

This exercise is good even if you still think you'll license the invention and just collect royalties. You will need the numbers to negotiate a licensing agreement realistically and to justify your position. **You can bet any licensee will have their MBAs do a thorough cost benefit analysis of adding your product to their line** versus adding someone else's (maybe even their own) to their product line.

They will ALWAYS proceed with the products for which there is the highest EXPECTED return AND which fit within their current resource constraints. Translated into inventor speak, that means that they may reject your idea/product because either they currently have other product opportunities with higher expected return (maybe just because there is no royalty cost) or their current resource constraints (dollars, labor, capacity, expertise, etc.) will be reached before they get to your invention. <u>That can be two strikes against you even if your invention will (eventually) be a winner.</u>

5. Ignoring industry rules. If you are not knowledgeable about the rules (both legal and informal) in the industry your invention is for, you should get help before expending too much effort on just doing things in whatever way comes to mind.

For example, an informal industry rule may be that catalog sheets (What are those?) NEVER disclose a suggested retail price. If yours does, whether well intended or not, you make your catalog sheet "special" in a very negative sense because now sales reps showing your catalog sheet have to treat it special in a way that is not likely to benefit themselves. Will that encourage them to tout your invention or relegate it to the back burner to be brought forward when asked?

This book does not follow all the standard industry rules for editing. It does try to follow the rules necessary for wide distribution. Whenever you come across a rule ask, "Who does this rule benefit?" If it is mostly "editors," or if it is you in some significant way, then you can make an intelligent decision. It's up to you; you do not have to follow industry rules for any of the packaging, price points, quality designations, etc. (unless required to by law).

After all, all of the non-legal (at least) industry rules probably developed over years because random people "invented" them one at a time in piecemeal fashion and they were adopted by others in the industry because they simplified interaction (i.e., lowered total cost) for the industry. You might just be the inventor that invents a new industry rule—but I will bet against you every time—I like near dang sure bets. On the other hand, while you can't change the rules you can make up a new game if that would be to your marketing benefit (see Marketing 101. Chapter 8).

6. Approaching tough markets. Sure, the entertainment industry is HUGE, all you have to do is get .00001% of it and you'll be fabulously wealthy. Of course you will have to wrest that tiny little percent from hundreds of thousands of existing, reasonably well-capitalized companies that already split that market into many pieces. You will also have to beat out all the other inventors trying to force themselves into that huge market too.

If you think inventors have it bad considering that maybe only 2% of all PATENTED inventions make money, you will no doubt be relieved to know that your lot in life is far from the hardest. I don't know what the hardest is but I do know that the music industry estimates that 3,000-5,000 songs are written for every 1 that is finally recorded, released for public sale, and achieves decent radio air play. (The odds for songs are actually improving since author/artists can now (due to computers and industry standards) create and release their own CDs to fans without ever worrying about air play. That still leaves the issue of profitability and ultimately the issue of whether the author/artists can wholly support themselves from their music.)

The best hint I can give you for breaking into the entertainment business is to listen to everyone you present your entertainment invention to. If better than 90% of them say "WOW" the first time they see it or understand it, you might have a pretty good chance at success.

7. Believing in improving the retailer's percentage. If you have any brains at all (and I assume you must because you've gotten this far) it has probably dawned on you that you likely have two markets you must satisfy, the end consumer and the retailer (if not other parts of the distribution channel).

Since your objective as manufacturer is to sell products to the channel and leave it up to the channel to sell to the consumer, you might just say "Why don't I just sweeten the deal for the retailer?" Instead of worrying about marketing to the consumer yourself you can let the retailer worry about it with their sweetened profits.

Just try presenting that idea to a retailer and see if they aren't still rolling on the floor with laughter after 5 minutes. All it takes for you to "sweeten" the deal for the retailer is to raise your "MSRP" (manufacturer's suggested retail price). How much value do you add to the product by doing that? None is the correct answer. Unless you offer your retailer an exclusive right to sell your product in some reasonable area AND your product is one that customers will want at the higher price there is no chance the scheme will work unless the retailer gets together with his competitors to fix the minimum sales price. A procedure which is clearly against the law.

In the late 60s and early 70s it was not uncommon for manufacturers (Magnavox, for example) to establish a "Fair-Trade" price below which they did not want their products sold. The idea being that selling below that price "cheapened" the brand name in the eyes of the consumer. For "Fair-Trade" to be possible, the state you wanted to sell in had to have a "Fair-Trade" law which allows a manufacturer to fix prices in a way that would otherwise violate Federal Anti-Trust laws.

For the most part I believe these manufacturer schemes supported by state legislation have proven to be too expensive to enforce and in general have been booed out of existence by the consuming public and the retailers. Clearly, a retailer would not sell a product below the manufacturer's minimum price unless he could still make a profit doing so. In other words, the manufacturer had a cheap product that was being palmed off as a more expensive product. Not good for repeat consumer sales or confidence.

The retailer will also laugh because they are keenly aware that there are two factors that determine what can be charged for a product (and neither one of them is MSRP). The first is what customers are willing to pay. Obviously some will be willing to pay more than others, but with the exception of very special products, the price range will not be too great relative to achieving a worthwhile level of sales for the retailer.

(At some point you may want to do price and demand elasticity experiments with your product but you may discover that it can vary considerably from city to city. As an obvious example, snowmobiles, regardless of price, just don't sell well in Florida. National retail chains crunch those kinds of numbers frequently.)

The second reason is that competitors also affect what the retailer can charge. All those office supply catalogs you receive or the number of grocery store inserts in the newspaper should make it clear that differences of even a few cents will make a difference as to which store a prospective customer will buy your product in.

Additionally, it is not improbable that after a retailer "buys" a product from you (or the channel) and watches it sit on the shelf without selling, they will ask for their money back. If you turn up your nose (or dig in your heels) and say "No way, sorry, your problem!" do you think you'll get a reputation as a manufacturer worth doing business with? No way! Will the channel give the money back and bring the product back to you? You betcha! Will they distribute more if they see a lot of products coming back? Will they increase that if you don't refund money for unsold stock? Fat chance.

James E. White

If you are at all familiar with the publisher/bookstore industry, you are aware that books are very often "sold" to the retailers under the agreement that the retailer can get their money back for returning the book (or often only the torn-off cover). Those kinds of agreements are rarer in other industries but they do exist. You and your product will not change the competitive marketplace.

8. Hiring hungry help. Often times I've heard or read that it is best to hire hungry consultants. That way you'll get the best rates and they will be willing to do what you want. I'm not so sure I wholly agree with that. Yes, you will definitely get the best RATES, but you may end up paying far more in the long run. A hungry consultant needs your paycheck AND they have the time to work exclusively for you and they often will.

Do you want someone who has the experience to cut right to the chase and give you more value for each dollar you spend with them or do you want someone who can take your fuzzy wishes and chase down all kinds of alleys that look interesting to them and that MIGHT prove relevant to you?

As previously noted in the Giant Paperclip (SuperClip) example, how an apparently cut and dried patent on "long arms and high-carbon steel" could cost $150,000 is beyond me, unless the inventors demanded challenges to knockoff producers that they HOPED would SCARE them off since the patent claims clearly were not infringed. Fat chance if there is big money on the table. Another possibility is that the inventor or the article author lumped <u>Trademark</u> costs in with what they called "patent lawyers' fees" just to give the reader a more dramatic number.

The SuperClip inventor did apparently go through several trademarks but whether they hoped to corner the market on product names (unlikely) or were just indecisive about what they wanted to ultimately call the SuperClip I don't know. Still, start to finish, legal help for a trademark should cost less than $1,000 plus the $325 application fee. I also don't know whether they might have spent some of that money on filing (losing) challenges against other giant paperclip manufacturer's trademark name filings or getting foreign patents that were just as worthless as their U.S. patent.

On hourly assistance agreements and on those requiring expenses you should always make certain that you clearly state the limits you place on your help. If that means saying "Do 2 hours of research on this issue and give me the results," then do it. If that means saying "Get the manufacturer's preference on doing this then get back to me for a decision," then say it. You are in charge—don't leave a vacuum for them to fill (or unguarded money to line their pockets with).

I think honest, competent help is a far better investment than shopping for the lowest rates. Make up your own mind.

When Is the Cheap Marketer Better?

For a homework exercise take a marketer who achieves $10 million in sales in the same time period another marketer achieves only $500,000 in sales. If you pay the one with the $10 million in sales a 15% commission and the other one an 8% commission, with which one do you have the most expenses and by how much? With which one do you have the most profit and by how much if your overhead expenses in the period were $1 million and your contribution margin 50%?

P.S. Just because I am a marketer, it does NOT mean that I put some kind of bias into this homework assignment. Really.

9. Waiting for financiers, marketers, etc., to appear. It will be a very rare case when Daddy Warbucks or SuperMarketer or whoever appears at your doorstep and volunteers to get and keep things moving till the big bucks start rolling into your pockets. Of course, 9 out of 10 or better of the people you approach for financial or marketing (or other help) and for which you don't agree to pay will turn you down.

That can hurt and certainly be very ego-deflating. What those people that turn you down know is that maybe one in a hundred product ideas are successful. They may not be able to tell the one from the rest, but they will certainly make the safe choice by turning down some inventor who can't seem to stay focused on the business issues help is needed on or who has no money and just wants their ego stroked.

You up your odds dramatically by first making sure your idea is worth pursuing and then by making things happen yourself. If you want to be successful then you have to do the work to achieve it. A guaranteed way to NEVER fail in the marketplace is to NEVER get a product out there in the first place. Zero at bats, zero strike outs.

I'll tell you now something from personal experience. A busy marketer or manufacturer will think it interesting when you volunteer that you have invented something. They will not, however, usually be induced out of

James E. White

curiosity or an insatiable drive for profits by that fact to take any action. The inventors who prepared and presented them with sound, formal, written propositions to which they were expected to respond are already tying up enough of their time—and probably generating profits.

10. Telling everyone who might help all about you. Again and again and again. Trust me, nobody cares. They are too busy wondering what you can do for them! And you can't do anything for them until you know what your next step is and ask for specific help in achieving it. They may disagree but at least you volunteered that you wanted help and that you were willing to take some action that involved them.

People in the invention industry will NOT be impressed that you achieved some award from some invention show that you paid $50 or more to attend. There are a lot of these shows and their biggest business is collecting entry fees and passing out "recognition" and award certificates. People will be even less impressed if the award is 5 or more years old and still nothing is happening with your invention. If you are in the "Inventors Hall of Fame," is it the real one (www.invent.org/book/index.html) or one of the other ones that have no "bricks and mortar"?

CHAPTER 7

Before You Enter the Marketplace

Well true, all of the work in the previous chapters happens before you enter the marketplace but what I am talking about here is the plethora of marketing items you will need to prepare to get your product selling and to keep it selling. Just advertisements, right? Well, no.

Trademarks

Do you know what sildenafil citrate is? Probably not. Do you know what Viagra is? Probably yes. They are the same thing. The first is the generic name of the drug and the second is the trademark for the drug by Pfizer, the exclusive manufacturer until the patent runs out (or they license other manufacturers to make it).

Do you need two names for your product? Maybe. If your product is truly new, you probably should find a generic way to describe it that your competitors (either now or after your patent expires) can use to describe similar products. Then you should also invent a trademarkable name that you use to identify your product by name.

You will need to be very careful to protect that name and to never use (or let others use) it as a generic name. If you fail to protect it, it will become generic and you will no longer have the exclusive right to use it. For example, "escalator" used to be a trademark of Haughton Elevator Company but has now become a generic name because they failed to protect it.

The details of getting a trademark are pretty straightforward. You can even read the *Trademark Manual of Examining Procedure* (TMEP) Second Edition, Revision 1.1 on the www.uspto.gov web site. The catch is selecting and getting a good trademark that isn't already taken and that the Patent and Trademark Office will register...then doing it again for foreign countries and hoping that your choice in the U.S. isn't offensive somewhere.

Trademarks can include color and even sounds and smells though the latter 2 are difficult to implement from a consumer perception/acceptance point of view. Trademarks can be symbols or illustrations as well as words but the most valuable ones are probably words. You can say words and type them into documents but how would you say ✪, for example, and distinguish it from ★ or ☆ and would everybody be able to include those symbols in a document? (Whatever happened to that guy formerly known as Prince?)

Symbols also must not be so generic they preclude competition in an industry. A drawing of an open brown cardboard box, for example, would not be registerable as a trademark for a cardboard box manufacturer because doing so might preclude other such manufacturers from publishing catalogs showing illustrations of their own wares. It might be registerable as a trademark for a jewelry item but I'm not sure it would become a valuable trademark.

There are, of course, some significant limitations to word trademarks too. The most basic thing to remember is that you register a trademark for a specific class or set of classes of products or services (you have to pay for each class you choose to include). In theory, at least, you could register COKE (the most recognized trademark in the world) for a class that The Coca-Cola Company has not registered it for. They have registered for about a dozen classes. In fact, out of 29 registrations including the word COKE there are 13 that do not belong to The Coca-Cola Company. Of course, being theoretically able to do something and being willing to argue against any opposition a well-funded giant might put in your way are two different things. The main ground they would have to argue against you on would probably be if your use of COKE "falsely suggested a connection" with them.

The USPTO, and the law, recognize essentially 4 levels on what is called the Distinctiveness/Descriptiveness Continuum. The first is "fanciful or arbitrary" and would include made-up words such as Pepsi and Kodak or words like Apple for computers. The second level is "suggestive" words such as Sno-Rake but you can easily get into trouble here because USPTO and court decisions have no clear rules.

Quick-Print and Bug Mist have been categorized at the next level which is "merely descriptive" and therefore could not be registered. If you can show 5 years of exclusive use (highly improbable for your new invention) and "recognition in the public's mind" of your trademark, you MAY be able to register "merely descriptive" words as having "acquired distinctiveness." The last level is "generic" words, such as "car" or "automobile," which cannot be registered, period (at least for their particular class).

There are also other special rules such as Surnames and deceptive, immoral, scandalous, and disparaging words (or designs) are not registerable.

So, after hours of thought and doodling you come up with the perfect fanciful or arbitrary word or words that you want to represent your invention in the minds of the public for all time. Or you come up with an exceptional suggestive word or phrase that the public is sure to understand but which is not merely descriptive or generic. Now what. First you research it to see if it is available. Be forewarned, this can be discouraging because you will often find you're too late. Some other @$&*&%# genius is ahead of you.

The first place to search is the USPTO *Trademark Acceptable Identification of Goods and Services Manual* online at www.uspto.gov/web/offices/tac/doc/gsmanual. Try searching a one word descriptor first (e.g., jewelry) and see if that gives you enough hits to clearly find your three-digit class number. If it doesn't then figure out other words or search strategies. Worst case is you have to retrieve about 800K of classification listing and read it. Next go to the Trademark database found at www.uspto.gov/tmdb/index.html.

Search your trademark word or words. Sometimes it's best to do one at a time but that can also give you an overwhelming list to wade through. Read through the list deciding which might conflict with yours and looking at them to see if they are for your class. If you complete that with no realistic conflicts, try the Canadian PTO site strategis.ic.gc.ca/sc_consu/trade-marks/engdoc/cover.html and do the same.

If you want to you can also go to Pipers at www.piperpat.co.nz/patoff.html and bounce to other trademark search sites in different countries. Everywhere you go, if you find something that you think MIGHT conflict, print it out.

If you still haven't found anything you are certain is a problem go to an Internet Yellow Pages or businesses directory for the U.S. and search for your word or words in company names. The one I typically use is www.worldnet.att.net/find/bgq.html by AT&T, but it is not the only one and I recently did some experimenting and concluded it DOES NOT provide true complete U.S. coverage. Be sure you use one that allows nationwide searching; some only allow single state and some (like GTE) seem to vary it erratically. Again try one word at a time then try a combination of your proposed trademark words to be sure you are getting reasonable results. And again print out any near hits.

Do the same for Canada canadayellowpages.com and wherever else you wish. I found a worldwide phone directory once and tried a few searches that

I had done in the U.S. and Canada but I was quite disappointed with the result so I didn't keep track of the site.

The caveat is use a trusted source—and then try a couple of other backup sources too. Someday the government will probably mandate instant and accurate phone database maintenance but until then the phone companies will want to play games with the information. If you detest their "monopolistic" behavior, remember that when you get your "patent monopoly" and want customers to genuinely like you and your product so they will buy again and again and recommend you to all their friends.

Still online you can use Register.com (www.register.com) or Network Solutions, Inc. (www.networksolutions.com) to check if your desired trademark or company name is already registered as a web domain name. You can also do an Internet search for "domain name registration" (without the quotes) and find a number of sites that will accomplish the search; some allow multiple simultaneous searches. If you choose to register a domain name be very careful about who you use or they will soak you (perhaps fraudulently) for extra fees beyond the registration fee of $70 for the first 2 years.

Unfortunately, the computerniks responsible for the Internet domain addresses are not (yet) using the power of computers intelligently. It should be possible for someone looking for a domain name to do a string search across all worldwide domain names to be sure they are selecting one that is not readily confusable. Alas, that is not yet possible.

There is also an ongoing debate at the World Intellectual Property Organization regarding what the rules should be on registration of similar domain names. For example, if you wanted to, should you be able to register kodak.net or kodak.uk and sort of masquerade as part of the Eastman Kodak Company? Should the site www.uspto.com, which is currently a "legitimate" site, be allowed to grab accidental accesses intended for the USPTO's ".gov" site? I, personally, wouldn't feel comfortable with any "intellectual property" attorney that associated themselves with such a site but you might see their domain name as a cute piece of intellectual work.

For the last stop in your own trademark search try the book *Brands and Their Companies* by Gale Research. You'll probably find the book at most large libraries with a decent business collection. Or you can try your luck online at the www.dialog.com or www.gale.com Internet sites. Both of these are pay sites and you'll probably have the easiest time getting in at Dialog.

If your choice for trademark is still "in the clear" there are still no guarantees because there could be a conflict with state, other foreign, or common law trademarks and even U.S. inactive applications and registrations. If you found some possible conflicts you might want to talk them over with your attorney and see if they are willing to guess at the probability of trouble.

Other ways to avoid trouble are to start over as many times as necessary to find a "clean" trademark or to contact the conflicting holder (if there is only 1) and see if you can work something out. **If they are not immediately on a win-win wavelength you should probably back out and start again with a different potential trademark.**

When you finally get to the point of proceeding with your trademark you'll probably want Thomson & Thomson to do a thorough U.S. (and perhaps international) search. You can contact them directly www.thomson-thomson.com or you can do it through your attorney. If you wander through their site you will figure out that some things you can do on your own through Internet sources (including their TRADEMARKSCAN® database via Dialog, www.dialog.com). For the $380.00 (U.S. search only) that I recently spent on a Thomson & Thomson search I know I could not beat their thoroughness or expertise. You will also want your attorney to review the document you get back and give you an opinion on any possible conflicts.

If everything still looks clean (or the risk appears acceptably low enough) you can follow the instructions at the USPTO site or have your attorney do the work. It will be more expensive but less nerve wracking to have your attorney do the paperwork (U.S. and international if you wish).

Keep in mind that for a trademark to be fully registered in the U.S. it must be put "in use," so if you believe your own searches are good enough you do not need to spend money on Thomson & Thomson or your attorney until after you have proven your product will sell and you reach a full GO decision. Even then the Thomson & Thomson search (or an equivalent one by another vendor) is not mandatory—if nothing is found, it just increases your confidence that no one will object to your trademark after you file a formal application.

Remember, initially you've only been using your trademark with the common-law ™ after it. You can still change it easily before larger scale marketing takes place.

The catch, of course (Why must there always be a catch?), is that if you don't apply for your trademark immediately, someone else may beat you to applying for the trademark and the USPTO may register it to them despite your objections and even despite evidence that you put it in use first. There are no guarantees, but the safest route, if you can afford it, is to apply with an "intent to use" application before you put the trademark in use, then to pay the extra fees when you actually do put it in use. Still, I think I'll take my chances and start using my trademark first (with a "™") then pay only once after I've got it in use.

The odds of theft are low (and thieves are usually easy to defeat in the long run) and the chances of another genius coming up with your trademark idea in your minimally protected 6-12 months of development or test marketing are usually slim.

Only after you have filed your formal trademark application and after it has gone through the application process and you have received notice that it has been "registered" can you use the ® after your trademark. In the U.S. this generally takes 8 months to 2 years.

If you register your trademark in another country and they also use the ® symbol then it is legal to sell your product so marked in the U.S. but your U.S. trademark rights are NOT protected. I can think of no valid reason not to do the U.S. paperwork. A possibly "valid" reason to try to skate by with a foreign registration even if your mark is turned down by the U.S. as "merely descriptive" is that you can later register it as having attained "distinctiveness."

At best, though, this will be a 10 year game you'll be playing (5 years to "acquire distinctiveness" and 5 to NOT lose to a challenge) and you will need the paid services of an attorney to play it. Other countries may have different registration procedures and notification marks than the U.S. so get help for international products.

Different USPTO Mandates

Incidentally, the mandates of the Patent side of the Patent and Trademark Office and the Trademark side are quite different.

The Patent side protects inventors (and thereby manufacturers) by giving them exclusive rights to their inventions (NOT THEIR IDEAS) for a period of time but at the same time is mandated to make inventors' work public for others to build on.

> The Trademark side's mandate is more to ensure (fair) competition and to provide a means for businesses to reassure customers that they are getting what they expect.

Why Can't I Buy a "Yes" Answer?

Another thing I'll mention here: Everyone wants a definitive "yes" answer to "Is it patentable?," "Is that trademark OK?," "Will people buy?" before proceeding with spending significant money such as a patent application or a trademark application or initial test market production. But you can't get it. Even after the patent is granted, the trademark okayed, and the test run sells out there is no guarantee. If the uncertainties of business don't mesh well with the clarity of mind of one who has perfected an invention, you should seriously consider engaging business partners or professional managers.

Marketing Item List

The following list identifies the types of things you will need to prepare before you consider your marketing and product fully launched. I have included some fundamentals about them but for considerably more info on many of these see Mr. Merrick's book.

Product Name—trademarkable, easy to spell, easy to pronounce

Generic Product Name—for you and others to use when talking about your product generically

Press Releases—who, what, when, where, why from most important to least important and in the style of news (You need a new one of these at least anytime something newsworthy happens, maybe as many as 1 a month.)

Display Advertising—for public, for trade publications (If Microsoft had to spend $200,000,000 dollars on the release of Windows 95 to get it noticed, how much will you need to spend?)

Classified Advertising—for public, for trade publications

Press Kit—envelope containing a folder with press release, photograph, and PR material (biography, catalog sheet, brochure, ad reprints, photocopies of press clippings)

Photo—glossy black and white (4" x 5" or 8" x 10") with caption taped to back; should tell story of product use and be high contrast

Retail Packaging—you may need more than one for different distribution channels; quality packaging counts to retail buyers but not so much to catalog buyers; anti-pilfering, easy to store and display

Distribution Packaging—must be aligned with your manufacturer pricing discount schedule (Boxing by dozens and pricing by hundreds won't cut it.)

Universal Product Code—for scanner recognition, consistent with stocking number (see page 150)

Inquiry Response Package (consumer)—catalog page or brochure, cover letter, sample if cheap consumable

Inquiry Response Package (retailer)—catalog page, distribution info, cover letter, perhaps sample

Sampling Package—for handing out at conferences or other group meetings; usually for inexpensive products; tailor product to audience if practical (e.g., silk screened or hot stamped conference or organization logo)

Catalog Sheet—for sales representatives and others in the distribution channel; should conform with industry norms

Price List (Manufacturers, i.e. yours)—for distributors, reps, etc.

Price List (Distributors)—suggested for distributors to give to retailers, for your use if you distribute, be wary of trying to be a one-product distributor, the extra paperwork costs born by your retailers may encourage them to pick a substitute product

Price List (Direct)—for end users that contact you directly

Business Cards

Stationary

Order Forms

Magazine Insert Cards

Statement Stuffers—like you get in your credit card bills

Direct Mail—envelope, letter, brochure, response mechanism (postcard, 800 number, etc.)

Catalog House offer—with suggested photo and text, pricing, drop ship offer

Radio

TV

Internet—with online, safe response

Sales Rep Agreement—often they will get 10 to 15% of your "manufacturers" price to distributors but it depends on the industry and can go below 5% and over 25%

Are you worn out just reading the list? Now do you begin to see why the people you wanted to sell your idea to may not be quite so eager to "take your idea and run with it"? Especially without any guarantee that it will sell or could be produced or would even work.

Most of the above you will obviously have to do yourself or pay professionals to do it for you, but you may be able to get some help on a couple of pieces, especially industry expectations for catalog sheets and price lists, from a manufacturer's rep. The easiest place to start looking for these is at the Manufacturers' Agents National Association web site at www.manaonline.org.

Marketing References List

Before starting through the above list and developing the individual items you might want to read the following materials. Everything on this list of materials is mentioned elsewhere in this book. The list is merely collected here for your convenience. This list, as is the full "Resources" list in Chapter 14, is alphabetized by title.

Creating Demand, Powerful Tips and Tactics for Marketing Your Product or Service Richard Ott, 1992

Do-It-Yourself Marketing Research George Edward Breen, 1977, 1998

Essentials of Marketing, E. McCarthy and W. Perreault, Jr., 7th edition

How to Develop Successful New Products, Jerry Patrick, 1997

Marketing 120, Principles of Marketing <u>webster.gtcc.cc.nc.us/vcampus/mkt120/index.htm</u>

Marketing Know-How, Your Guide to the Best Marketing Tools and Sources Peter Forancese, 1996

Marketing Your Invention Thomas E. Mosley, Jr., 1997

Stand Alone, Inventor! And Make Money With Your New Product Ideas! Robert G. Merrick, 1997-1998

CHAPTER 8

Marketing 101 (Abridged)

The first rule of marketing I learned back in business school in the 1970s was:

"Find a need and fill it."

It's a pretty simple rule but often a lot easier to say than to put into practice. A person with a new hammer, they say, sees everything as a nail. A person with a new invention idea often sees everyone else as having a need for the invention.

Bob Merrick's version of the rule is "Marketing is producing what customers will want" (pg. 73).

I prefer to narrow that even further and work only with products and services that are "in the best interests of the customer." But this gets into a huge gray area really fast. For example you might include alcohol or tobacco within the confines of "best interests of the customer" while someone else would exclude it. **Is your invention truly in the best interests of the customer and is it something people want (or need)? Answer "yes" and your invention passes the hardest hurdle in marketing. Answer "no" and you should very seriously consider having a new invention idea.**

In a nutshell that is the important part of marketing that I want you to get out of this book. Perhaps in the future I will write another book about the size of this one that deals exclusively with marketing—as applied AFTER production starts (or is nearly ready) on a product that is <u>proven to sell</u>. I can already tell you what the basic premise of that book, if I get to it, will be. **The bulk of your marketing efforts will likely be to the market channel and NOT to the consumer.**

Doug Taeckens, the president of a household products manufacturing firm with about 50 years of invention experience, defines marketing as the following:

"Marketing is the art of getting the right product to the right people at the right price through the right channels by the right promotion."

James E. White

Read that very carefully and note how carefully the words were chosen and the sequence of the elements. It's product then people then channel then promotion (not necessarily advertising). The promotion can increase awareness and perhaps enhance desirability but it cannot overcome problems with product, people, or price.

Since I know you won't have dropped everything to read the books in the list I included at the end of the last chapter, I'll help you out some by providing a few pages on some of the basics. The next chapter, which warrants a book itself, will touch on advertising.

For the most part this book assumes that your market channel will be the conventional one, i.e., manufacturer to distributor to retailer to consumer. That is by no means the only possible market channel. There are market channels direct to consumers, direct to business, to specific employee types (i.e., purchasing agents), direct to retailer, special order only, drop ship, factory outlets, etc. Don't get too concerned about the market channel until after you really have a product to sell.

Rest assured, YOU will not change the market and neither will I. Everyone will not ever own a Mercedes-Benz—even if they can afford it—because not everybody is part of the luxury car market. Not everyone will own a Hula Hoop because that is not the right kind of entertainment device for a large segment of the market.

I will not expect to get even 1% of the inventors' marketing market because, A) I suspect that 95% of all inventors are so risk averse they will never get their invention to market and, B) I'm keenly aware I have competitors. But then, I'll be happy to get a handful of inventors' products to market for inventors who are serious about generating income (and perhaps creating wealth) as inventors. (Hint: These are not likely to be individuals who are just interested in selling their idea to me or someone else.)

Determining Market Size

"This idea is fantastic! Everybody can use this product."

Marketers, and probably patent attorneys, hear that a lot. Even IF it's true, the truth is that NOT EVERYBODY will buy one. Everybody conceivably can use a Rolls-Royce for transportation, but I don't see one in any of my neighbors' driveways yet. (Or mine either in case you were wondering.) Earplugs cost about 69 cents a pair and everybody could use a pair from time to time. Do they though? I still have never bought a pair even though I've thought of it from time to time. I've never owned an umbrella in my life (but I've shared the protection of one on rare occasions).

You too can probably create an enormous list of things you COULD use but never have. Very few supermarket shoppers use more than 15% of the products available there. I would be amazed to find a grocer, who presumably knows everything that is there, who had ever used 25% of the products (unless they systematically tested things). By now you should realize that just because "everybody could use your invention," doesn't mean that "everybody will buy one." **It's time to be more realistic.**

Some basics for determining market size are presented in the following pages. This is an essential task for doing any of the suggested profitability calculations in the previous chapters. The very first thing you need to do is define who your market is. Just because you have a fishing lure your market is not all U.S. (or worldwide) males. It is most certainly not exclusively males and it is certainly not me. If you think about it, it probably won't even be all fishers because some are going after different kinds of fish in different environments and with different tackle than your lure is suitable for.

The game here is to be as exact as you can. You can be exact in several different markets if applicable, but be exact. Of course being exact cuts the numbers down but if your projections show a profit you can be pretty confident. If you inflate the numbers and show a marginal profit, you are a financial disaster waiting to happen.

There is no right or wrong place to start looking for numbers because it depends entirely on the target market. The following are presented in no particular order.

Census Numbers

U.S. Census Bureau information is available at www.census.gov. It is cute but of no particular value that the total U.S. population is updated every 5 minutes at this site (270 plus million). Let me walk you through one search I recently did and you will get some idea of what you can do. My invention is for use with necklaces and primarily to be worn by females. From the main page I selected "(People) Estimates" then "National" then "Annual Population Estimates by Age Group and Sex" then "Population Data." I'm only interested in the most current data so I look at the last column and see that the total population is 270,933,000 which I write down because I know I will need it for later calculations. You should write that number down too since you will need it later for computing ratios, etc. Then I scrolled down to the section on FEMALES.

At the bottom of the section is "Special age categories" starting with 16 years and over—well, 14-year-olds might want my invention also but I'll be conservative and not count them. The number for 16 and over for 1998 is 107,939 and if we look at the notes at the top of the section we see that all numbers are in thousands so the total for 1998 of U.S. females 16 and over is 107,939,000. A significant market—but I won't forget that not all women wear necklaces suitable for my invention and not all of those that do will want the invention. I could have also looked up data by sex, race and Hispanic origin if those were relevant.

Just for sport I also looked to see how many prospects there were in my test market area. I backed out to the screen in which I originally chose "National" and then chose "Metropolitan Area" then "Metropolitan Area" again to get to the full list of U.S. Metropolitan Statistical Areas (MSA). Since these are typically by major city I looked down for Lansing and found "Lansing-East Lansing MI" including parts of Clinton, Eaton, and Ingham Counties.

I could have searched the page with **Edit|Find** but chose instead just to scroll down scanning the list and making certain that I kept my eye on the correct column to avoid looking at the wrong alphabetized grouping. There are 447,538 people in my test market area. To figure out the number of females 16 and over I used the numbers I got previously in the following ratio formula.

$$\frac{\substack{\text{US Females 16}\\\text{and over}\\107{,}939{,}000}}{\substack{\text{US Total}\\\text{Population}\\270{,}933{,}000}} \times \substack{\text{Lansing MSA}\\\text{Population}\\447{,}538} = \substack{\text{Lansing MSA}\\\text{Females 16 \& over}\\178{,}298}$$

A right nice test market area. I also note that population growth for 1990 to 1996 was 3.4% in the last column or 14,854 in the next-to-last column. I could use those numbers in doing business plan projections 5 years out. But, you say, there is no guarantee that growth will continue just like that and there isn't even any certain reason to believe the female population of the Lansing area matches that of the U.S. as a whole. You are absolutely right! I would suggest, however, that if your profitability depends on a 10% difference in the data I'd drop that invention idea immediately—odds are you are stretching all numbers and believing in 100% penetration. Call me today about the land I have available in Florida, please.

Next I went back out to the main Census Bureau page and picked "Income" because I thought that might impact necklace wearing. I could have picked from "Housing" or other categories if I thought they might be relevant. For my purposes I picked the "Ferret Data Extraction and Review Tool" after which I had to enter my e-mail address in case my request would have to be queued and I would have to pick the results up later. I clicked "Continue" then picked the "Survey of Income and Program Participation -1996" and clicked the "Continue" button at the bottom. I also noticed there was information on Job Tenure, Race, Voting and Registration, Health, and other topics which are not of interest for this search.

I tried looking at tables already created first and found that "Table 6: Monthly Earnings by Sex, Race, Work Experience" looked like it might be useful. The average monthly earned income of working females is about $1,400. Since my product will be priced under $5 I figure they could at least afford to buy one or two a year. That suspicion was confirmed when I looked at a previously issued 1996 graph that showed median household income at about $35,000 annually.

A little further rummaging on the Income page showed "Earnings by Occupation and Education: 1990" which jumps out to the Oregon State University site (govinfo.kerr.orst.edu) which also has Import Export data, Agricultural, School District and other data that might be useful to look at for some other project. For now I'll stick with "Earnings by Occupation," "United States," "All persons 18 years and over in the experienced civilian labor force who worked with earnings in 1989." Looking at the bottom of the report, there are 52,298,056 full or part time female employees in the U.S.

Also at the bottom I selected "Mean annual earnings" and clicked "Get selected report" to find that the mean annual earnings for females was $16,593 (which agrees with the $1,400/mo. x 12 = $16,800 from above). Looking a little closer I also noted that for males the mean annual income was $29,984. Now I see the light: since males have higher incomes, the trick is to get males to buy my product as a gift for females. Sales to females will just be gravy. (That lump in my cheek? Huh? Is there a lump there? I have no idea what it might be.)

Well wait a minute, while I'm here why don't I see if I can find out who might buy the Fancy Wheel Lock Nut Removal tool. In this screen we can search for an occupation, say "Mechanic", then choose "Automobile mechanics, except apprentices (505)" from those occupational categories found. The table shows that the total number of these employees in the U.S. having year round full time employment was 668,391. There were 910,479 including part time and the average annual full time earnings were $23,793 in 1990. Even if one is sold to every mechanic in the country there will never be a million tools sold. Well guys like tools; we could target guys in general instead of just mechanics. Hello, are you listening?

To continue looking for potential tool sales I went back to the Census Bureau home page and selected "(Business) 1997 Economic Census." When I tried to look at 1997 data under "Latest Results" I discovered two things. The first was that the U.S. is now caught in the transition from SIC codes to NAICS. The North American Industry Classification System is replacing the U.S. Standard Industrial Classification (SIC) system. NAICS was developed jointly by the U.S., Canada, and Mexico.

The second thing I discovered was that the 1997 data (as of May '99) was far from complete, and won't be complete till sometime in 2002, so I can't find what I want there yet. I looked in 1992 and tried "Service Industries (taxable firms) National" and found "Automotive repair, services, and parking" then picked "75" in the SIC column and found that the 753 classification, "Automotive repair shops," identifies 128,738 U.S. locations employing 519,903 people (which probably are not all mechanics either) and doing about $39 billion worth of business.

Why the discrepancy in the number of auto mechanics between the People data and the Business data? I don't know for sure but I would suspect that there are a large number of fleet mechanics working for such things as taxi, police, and other businesses that are not classified as Auto Repair and that 43,000 "New and used car dealers" employing 920,000 people include a fair number of auto mechanics. The fleet shops would be hardest to target but are unlikely to need the tool anyway.

In other words, THINK about the numbers and what they really represent and then put that in the context of your invention.

If you looked closely at the 753x classification you would also find that not all shops are likely to need the tool because they do muffler or glass or other specialty work. In fact it appears that only about 70,000 shops employing around 350,000 people plus 40,000 auto dealerships employing an unknown number of mechanics might need the tool. Plus the tool would also be of interest to wheel and auto thieves.

The bottom line then is, a totally saturated market would probably need a max of about 200,000 of the tool (about 2 for each shop and not counting thieves) unless, of course, new locking mechanisms come on the market which make the tool essentially obsolete. Actually, our data search isn't really able to tell us how many will sell, but it does clearly give us some maximum numbers which it would be ridiculous to assume can be achieved.

After 10 years very few products will have more than 10% of the total target market so start by assuming that 10% of your total target market is your total projected market. Even with that said, my bias is not to worry too much about the total potential but just to worry about how much you have to sell to break even (i.e., until payback) and start making a profit. If you can reasonably predict your total startup costs, estimate your ongoing overhead costs, and clearly identify your average or approximate manufacturer's contribution margin in dollars, the computation is straightforward if tedious.

Compute Your Invention's Breakeven Sales

First you'll have to make some projected guesses about the number of units that will sell in the next 3 or more periods. I'll assume sales numbers are for each year but you might want to break things down by month or quarter.

On a sheet of paper or in a spreadsheet program start 3 accumulator columns, one for Units Sold, one for Net Receipts, and one for Indirect Costs. In the Units Sold column enter the number of units you expect to sell in the first period. In the Net Receipts column enter the number of units times the contribution margin in dollars. In the Indirect Costs column enter the startup costs and the overhead costs as negative numbers for the period then sum them.

Draw a line across all three columns below your current entries. "Total" the units sold through the period and enter it just below the line in the Units Sold column. Add the number in the Net Receipts column to the sum in the Indirect Costs column and enter it below the line in the Indirect Costs column being careful to enter the sign correctly. That represents your "startup costs" for the second period.

Continue with your projected sales and overhead costs for the next period doing the same steps as before. When you finally enter a positive number as the next period's "startup costs" you can count that as the period in which you start to see real profits. If the number at the bottom of the Units Sold column at that point is substantially less than 1% of your total projected market (or .001 times total target market) and all other factors are still favorable, you stand a good chance of surviving and prospering. Keep in mind that the IRS definition of profits is based on amortization of startup and some R&D costs so you can pay income taxes in years when cash flows were (or still are) negative.

Jewelry Invention Payback Estimates

Manufacturer's Contribution = $1.00/Unit
Startup Costs = $3,000
Target Market = 100,000,000
Projected Market (10 years total) = 10,000,000

Period	Units Sold	Net Receipts	Indirect Costs
			-$3,000
Year 1	600	$600	-$20,000
			=-$23,000
	600		-$22,400
Year 2	30,000	$30,000	-$6,000
			=-$28,400
	30,600		$1,600
Year 3	500,000	$500,000	-$50,000
			=-$48,400
	536,000		$451,600

Payback after 2 years with 0.36% of the Projected Market

If you think about the above table you will note that I have projected a large market and a small contribution margin but I still won't get a decent income till after 2 years of beans and hotdogs. If your market is considerably smaller you will need to boost your contribution margin significantly to get a decent income. You should also be aware that my projections are based on a slow rollout and growth from actual cash flow. If I were to accept financing I would increase my costs but could probably boost growth and short term profits. If I move into international markets starting in the 3rd year I won't get to pocket all the "profits" shown in my projection either.

Mailing List Numbers

Another source of numbers is mailing list counts. These are also often available online. An amazing Internet service to find lists in the UK, the U.S. and internationally has been set up by List Link. If you go to their site, www.list-link.com you can rummage around and perhaps find your own list. At the time of this writing you could get in for 15 days free or get a month of access for $85 after that. Your initial registration must be processed and a response sent to your e-mail address before you can actually access the site. I don't have enough experience with them to know how satisfactory the results might be.

A Secret Mailing-List Industry Rule

NOTE: lists are generally sold on price per thousand with a 5,000 minimum EVEN IF IT ISN'T EXPLICITLY STATED in what you are reading. Even though that is not a 100% rule it is one you will be expected to "know" because it helps the mailing list industry distinguish "probably never buyers" from "serious buyers." Always clarify before agreeing to an order. At this point, of course, you are just looking for some numbers, but in the future you might use these sites to order mailing lists that specifically target your prospects.

Other online list sources include:

www.edithroman.com Edith Roman Associates, Inc. has mainly U.S. list information but requires that you logon for free before it lets you in. You will get a password and ID back in your e-mail in a day or so. Some parts of the site are only accessible after you call them. I looked at the compiled lists with numbers and found 7,032 as the number of "Jewelers, Wholesale" and 44,276 as "Jewelers, Retail" plus about 2000 miscellaneous other jewelry-related businesses. These numbers are reasonably in agreement with the same categories found in the census data.

www.directmedia.com Acxiom Direct Media provides direct access to rate cards with numbers of addresses. Search is by keyword so you may have to try a few to hit the right word that they have chosen for finding the kind of people/businesses you are interested in.

www.dunhills.com Dunhill International List Company. Pick the Datacards link and you will get a very disappointing list (to me at least) of datacards but the datacards do give the numeric info you may need to estimate a market size. They also give the base rate per thousand (M) and surcharge per thousand for various categories such as age, sex, and state. Request their mailing list catalog and you'll get a treasure-trove of numbers.

www.polk.com Polk Direct. This is an excellent site for consumer lists. Their rate "cards" are in pdf format (essentially it is just their print catalog pages online) so be aware of that before you go searching because you'll need the Adobe Acrobat reader (www.adobe.com/prodindex/acrobat/readstep. html#reader). They are very good at providing numbers such as "Home Workshop/Do-It-Yourself" = 8,906,000 if you pick the right rate card. Other rate cards have no hard numbers other than the premium you pay per thousand for refining other lists by the card's category.

There is a U.S. service called Marketing Information Network (mIn) which claims to have a list of almost every U.S. and Canadian list and weird media. They give this info to 400 associated list brokers. The list of the list brokers can be had from lists@minokc.com or fax (405) 575-1055. The service to brokers includes a toll free number with a database searchable using keywords.

www.amlist.com American List Counsel provides access to a lot of good articles on direct mail marketing which could be highly useful if that becomes your preferred marketing channel.

If doing your own searching for a relevant list turns up nothing or for some reason turns out to be too difficult and you would like help, call a list broker. Look in your phone book under mailing lists. List brokers are paid by the list owners (when orders are placed through the broker) so are free to you when you are just asking questions, but you can wear their patience pretty thin pretty fast if it looks like you won't be doing any ordering.

Why mention all those? To give you some idea of just how big a business mailing lists are and how narrowly you can focus on your target market through buying mailing lists. If your product lends itself to direct mail a good list can be a gold mine.

Be forewarned though, direct mail is a very tough market. "Good" response rates can be as low as .1 or .2 percent and some products don't get solid bites till the third or fourth repeat mailing. A 1% return would be considered very good and 10% would be astronomical. "WOW!" type products to precisely correct lists have been known to hit 40% responses but the decline in the sales curve can be just as rapid.

Factor the cost of repeats and rejected mailings into your profit projections. DO NOT do what I've seen some people do when making estimates. They figure the 11-cent bulk rate mailing cost and 50-cent printing and stuffing and addressing costs of sending a mailing to a customer will easily be covered by the $9.95 price of the cheap product that cost them 50 cents. That would only be true if you got nearly 100% responses and that will NEVER happen. The correct formula for marketing cost per product is total mailing (or marketing) costs divided by number of units sold.

$$\frac{\text{Marketing Cost}}{\text{per Unit}} = \frac{\text{Total Marketing Costs}}{\text{Number of Units Sold}}$$

Do test mailings of 5 or 10,000 first to determine what the number of orders per thousand mailed is so you can determine whether a nationwide mailing program will even have a chance of netting you a profit. For truly new products to sell well through the direct market channel usually requires, 1) that the problem be a significant one (meaning they will pay good money) for many people, 2) that you can sell in the price ballpark of your competition or have a clearly superior product, and 3) that you have clear outside expertise backing up your claims.

If you have a cheap product you may get a lot of people to try it once but if it does not obviously live up to your claims, the repeats, referrals, and "buzz" won't be there. ("Buzz" is what happens when "everybody" is talking about your product, when it's important enough to show up on the "right" TV situation comedies, etc. The Rubik's Cube had buzz.)

Conservative Market Penetration Estimating

What are good ways to determine "conservative" volumes and market penetration rates even though we don't know if the product will sell yet? Unfortunately there are no good guaranteed ways. Take our friend Thomas Edison. After spending at least $10,000—in 1879 dollars when cheap labor cost 7 cents an hour—to perfect the lightbulb it was an instant hit, right?

Wrong, some $200,000 plus later it was commercialized and 3,144 light bulbs had been sold to 203 customers by sometime in 1882. By 1889, 10 years after the patent, there were only 710 customers. The problem was that electricity and its support infrastructure cost too much. Ten more years later, after electricity's costs had come down, there were 3 million customers and all the basic light bulb patents had expired.

In an April 13, 1998, Newsweek article (pg. 14), "Reeling In the Years," the approximate time, from invention until 25% of the U.S. population had the household use of the invention, was provided in the following table:

Years Between Invention and Use by 25% of U.S. Population

Date	Invention	Years Till Mass Use
1873	Electricity	46
1876	Telephone	35
1886	Gasoline Automobile	55
1906	Radio	22
1926	Television	26
1953	Microwave Oven	30
1975	Personal Computer	16
1983	Mobile Phone	13
1991	The World Wide Web	7

Did you note that only 3 of them hit the 25% mark before a 20 year patent would have expired IF it could be gotten. Did you also note that of those 3, NONE have competition limiting patents. While working on this book I also talked to the president of a company that invented a roller for removing lint from clothing. The invention occurred in 1956 and as of 1998 they were in only ⅓ of all households. (My household happens to use two different BFH type solutions.) It is also true that even "universally" accepted products are not in all households. Only 98% of all U.S. households have televisions and only 80% have VCRs.

Looking a little more closely at the percent of U.S. homes with personal computers (as reported at www.ntia.doc.gov/ntiahome/fallingthru.html) we find the following:

Percent of U.S. Homes with Personal Computers

1985:	13.6%	1990:	25.4%
1986:	16.6%	1991:	26.6%
1987:	19.4%	1992:	29.6%
1988:	22.4%	1993:	32.7%
1989:	24.0%		

I also think I remember seeing somewhere that in late 1998 or in 1999 the percent of U.S. households with personal computers broke 50%. There is not enough data there to plot the whole classic market penetration curve as shown in the generic Product Life Cycle graph on the next page.

The percentages, of course, are simplified averages, but they should be a good place to start from before you have real sales data with which to adjust them a year or two from now. The most important thing you need to take away from this graph is very subtle, that is those are percentages of YOUR PROBABLE MARKET, NOT YOUR TARGET POPULATION. Very, very few products approach sales at the total target population level.

However, some products just keep on selling till they exceed the target population level by factors of 2 or 3 or more. Can you think of one or two? Those are the kinds of products you want to invent and sell. A new "classic" example is the telephone. How many do you have in your household versus how many do you need? Of course the basic telephone patents expired long (nearly 100 years) before telephone sales exceeded the target market size. In fact, telephones only significantly surpassed the target population level in the last 20 years—after deregulation of the telephone industry.

Some other things to note about the graph. You'll note that a few things are omitted. There are no time periods or dollar indicator lines and neither the target population numbers nor the actual market numbers are known. You'll also see that the curves are for the Industry as a whole, not just your firm. With proper business decisions and marketing I hope you'll dominate your industry, or at least maintain a large enough share that your sales follow the curve. The bigger your sales and the longer the product life cycle the more

Product Life Cycle

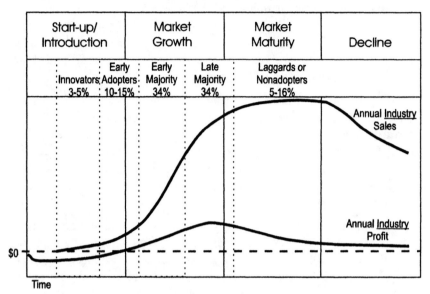

Start-up/ Introduction		Market Growth		Market Maturity	Decline
	Early	Early	Late	Laggards or	
Innovators	Adopters	Majority	Majority	Nonadopters	
3-5%	10-15%	34%	34%	5-16%	

Annual <u>Industry</u> Sales

Annual <u>Industry</u> Profit

$0

Time

likely you will have competitors. Obvious, of course. How do you get a handle on filling in the missing numbers? First get your target population size as described earlier in this chapter.

Sales Frequency, Product Life, & Competitor Availability Estimates

Now estimate, in years, the replacement frequency of your product. For products that are "never" replaced (a gold chain necklace or this year's hot Christmas fad toy, for example), always use 20 unless additional purchases are probable; then use what you estimate number of years between additional purchases to be. Gold necklaces, by the way, after over 4,000 years are still in their "Market Growth" phase. A fad Christmas toy might look like a tiny nail standing on its head by comparison. A weekly consumable would have a year's frequency number of 0.02 (1 week divided by 52 weeks). For electronic products (such as TVs, stereos, calculators) a typical number might be 8 years between replacement.

Next realistically estimate the number of years your product will generically stay on the market up to a maximum of 20. If your product is a refrigerator or television type product, 20 years would be appropriate. Ignore the fact that any particular model of refrigerator or televison might only have a product sales life of 3 years. A fad Christmas toy should probably get a number of somewhere between 1 and 2. A good way to get a handle on this is to ask, "How long has the problem existed?" and "How generally replaceable is it technologically?"

If the problem predates dust then you'll lean toward a high number. If your invention is easily replaced technologically then you should lean toward a low number. The problem of childhood boredom, and thus the need for toys, came shortly after dust but any particular toy is readily technologically replaceable so give a toy a 3 or 4. The need for a fancy-lug-nut-for-fancy-wheels removal tool is for a recent problem and it is highly probable that it is technologically easily replaced. Give it a 1 or 2.

One more estimate. How many years have competitive products been on the market, again with a maximum of 20? Toys would get a 20, fancy wheel nut removal tools, if there are no competitive products, would get a 0. Now you have 4 numbers: target population, replacement frequency, generic life, and competitive success. Plug those numbers, along with product price and contribution margin, into the appropriate places in the following formulas for at least 5 time periods (t) and plot them on an x-y graph with time as the x dimension.

Sales Projection Formulas

$$t_n = \sum_{1 \Rightarrow n} Target\_Population^{2.4} \times \frac{\sqrt[12]{Product\_Price}}{First\_Period\_Sales}$$

$$\$y_n = \int_{1 \Rightarrow n} Replacement\_Frequency - \left[2 \times \frac{Contribution\_Margin^2}{Generic\_Life^3} \right]$$

Your result should look something like a hyperbolic tangent function curve or similar to the first half of the Product Life Cycle curve a few pages back. Uh..., what? You can't quite see how to do the homework assignment? There is something you don't quite understand about the use of my formulas? Gotcha! Yes those numbers I asked you to generate to (presumably) plug into

James E. White

the formula are all ones you need to consider. Unfortunately there is no PROVEN model or formula that can be reasonably used. The Bass Model, by Frank Bass in 1969, is the classic and probably still among the best. The model involves an exponential pattern in which the number of purchases in period t_n depends on the number of purchases in periods t_1, t_2, t_3, ...t_{n-1}. In other words, if there is an error in your first guess, the error in the results gets worse faster. Bass and others also caution that the model is highly dependent on your guesses and its **correctness will also be subject to effects from the general economy while your product is on the market**.

A few years ago a book came out that pulled together papers on all the product sales forecasting models available at the time. All models were concluded to be quite accurate by their proponents BUT NONE PROVIDED ANY RIGOROUS EVIDENCE to back up the claim. Also none provided a useable mathematical description of the model used but we are assured it is properly embedded in the computer code. There are also many Internet sites by marketing firms touting models.

In the final analysis, most of the models are proprietary and they are subject to the "skilled" input of the firms that own them. It was surprising to see how many "academic" models were only available through consulting firms owned by the academic that invented them. Most tout "high" accuracy rates, but I suspect that a "good" accuracy rate is relatively easily obtained if the projected estimates are reasonably conservative and the manager responsible for product sales understands that if he gets sales to come in at the predicted level (plus or minus about 5%) his boss will be very happy with the predictions and the manager. Hitting the targets is good, not because it validates the models, but because it makes managing business budgets and business infrastructure growth easier, cheaper, and less subject to wild fluctuations.

If it is a financier or potential licensee that you need to impress, by all means use a model of some kind. Otherwise, use your best guesses and back it up with what you learned from focus group, mall intervention, or individual contacts and making the best use you can of the three "years" numbers you just estimated.

For a totally new product, the IBM 650 computer back when computers were a business unknown, the IBM market forecaster's best estimates for future sales were 5 to 500. This is after hours of focus group and one-on-one sessions with potential buyers. The prospective buyers were interested but would not commit until they saw the product in action. As it turned out sales wildly exceeded even the most optimistic estimate—BUT PROFITS WERE PROJECTED (and occurred) even at mid-level conservative estimates.

In the book *Danger, Marketing Researcher at Work* Terry Haller, in 1983, concluded that "90% of all [market] research is so seriously flawed as to be of questionable value." He made that estimate based on years of business experience in marketing at Proctor & Gamble, R. J. Reynolds, and Quaker Oats and from a study of 45 published market research studies of which all but 4 had obvious, serious, logic flaws.

I've seen no recent evidence to convince me his findings don't still hold. To you that means that IF YOU DO YOUR OWN market research, while being realistic and conservative, your research is just as likely to be as good as the professionals'. All it will lack is the fancy binder and the confident assertion made as the final payment check is received.

Other Market Statistics Sources

The first other place to look is the *Statistical Abstract of the United States* which can be found in annual volumes at many libraries or at www.doc.gov. At the web site choose "Site Search" then in the search string box enter the following string WITH THE QUOTES: "statistical abstract" to locate pages with that word pair. If you omit the quotes you will get thousands of useless hits to wade through. From the hits you'll probably be okay just selecting the first one for the *Statistical Abstract of the United States* but you may have to try more than one to get to the correct page.

As of this writing the current page was www.census.gov/prod/3/98pubs/ 98statab/cc98stab.htm but this is very likely to change. The correct page will contain a table showing "Section" and "Year" columns at the Census Bureau site. The Census Bureau, the Patent Office, and the National Institute of Science and Technology, among other government offices, are all part of the Department of Commerce (DOC). Now feel free to wander to your heart's content but be aware that you will be retrieving documents in pdf format and therefore you must have the Adobe Acrobat reader to view them.

As a couple of examples of what you can find there I selected "Section 14. Income, Expenditures, and Wealth" in the 1998 column and downloaded the pdf file, then just kept looking to find table "No. 723. Personal Consumption Expenditures..." which I took a closer look at. The table showed that for 1996, "Jewelry and watches" sold to the tune of 41.6 billion dollars. Let's see, .01 percent of that for my jewelry invention and I'll get...well, quite comfortable. In another table I also found that "Jewelry stores" only sold $20.1 billion worth of merchandise... "rats," I won't be quite so comfortable.

From the Statistical Abstracts page I also downloaded the "Guide to sources of statistics" which occurs toward the bottom of the table. In looking at it there is a "Retail and Wholesale Trade" section which lists the *Annual Benchmark Report for Retail Trade*. To find that I went back and did another search of the Department of Commerce site with "Annual Benchmark Report for Retail Trade" in quotes as my search string. I had to look at a couple of the pages that were found before actually getting the report but when I did finally retrieve it I found the following table (Numbered as Table 5 in the original):

Estimated Gross Margin as Percent

SIC code	Kind of business	1995	1996
	Retail trade, total	32.0	31.6
	Total (excl. auto group)	36.0	35.9
	Durable goods, total	26.6	25.9
52	Building mat., hardware, garden supply, and mobile home dealers	31.1	31.0
521,3	Building mat. and supply stores.	29.0	29.2
525	Hardware stores	37.2	37.6
55 ex. 554	Automotive dealers	19.2	18.3
551,2,5, 6,7,9	Motor vehicle and miscellaneous automotive dealers	17.9	16.9
553	Auto and home supply stores	38.5	39.1
57	Furniture, home furnishings, and equipment stores	36.1	35.3
571	Furniture and home furnishings	43.9	44.4
5722,31,4	Household appliance, radio, TV, and computer stores	27.5	25.3
	Nondurable goods, total	35.6	35.6
53	General merchandise group stores	29.6	29.2
531	Dept. stores (ex. leased depts.)	32.8	32.4
533	Variety stores	34.8	35.1
539	Misc. general mdse. stores	15.7	14.6
54	Food stores	26.7	26.7

541	Grocery stores	25.2	25.3
554	Gasoline service stations	22.7	22.3
56	Apparel and accessory stores	40.7	41.6
561	Men's and boys' clothing and furnishings stores	43.2	45.3
562,3	Women's clothing, specialty stores	42.4	44.4
566	Shoe stores	44.6	44.0
58	Eating and drinking places	64.9	65.0
591	Drug and proprietary stores	26.6	27.4
592	Liquor stores	28.2	28.9
53,56,57, 594	GAF(1)	34.6	34.3

"Gross Margin" is defined as "...total sales less cost of goods sold" and "GAF" essentially means department-store-type-merchandise stores.

From this table I learn that the typical margin a retailer in the jewelry business would expect to get above their cost for my invention is at least 40% as shown by SICs 56 and 562,3. That means I should try to have my product's wholesale price at about 50% of what customers will be willing to pay for it. (As noted in the Pitfalls chapter, that is NOT the same as MSRP.) If it is a really fast-selling product, I might be able to get away with the "average" of 35% but I'm probably better off using the higher number in my business plan calculations. The higher margin will be especially persuasive to retailers during product launch but will also encourage retention if sales are somewhat slower than the retailer's average product and depending on how much space and other resources my product takes up.

The table only shows averages by general type of store. It often occurs that some types of product lines provide a higher retail margin than others. Don't hesitate to ask several people in the industry. The general rule is that the larger or more expensive or slow selling the item the higher the margin required for the retailer to be satisfied with carrying it.

A couple of "gotchas" to be aware of relative to the *Statistical Abstract of the United States*: First, not all of the sources that it cites are findable via a search on the DOC site. Some can be found online using general search engines but others will require visiting a large library or contacting the source organization. If the abstracts contain some appropriate numbers, why would you want to go to the original source?

James E. White

The quoted numbers are often just the highest level of the statistics being abstracted. If you feel you can get by with that level, fine. However, lower level statistics may have significant numeric breakdowns that can be of use to you but that are not quoted in *Statistical Abstract*. For example, the abstract may have "Jewelry" while the source has "necklaces, bracelets, rings, pendents, settings," etc., individually itemized.

Second, if you don't find the statistic you are looking for in the most current "issue," don't hesitate to look through previous years. I would even encourage you to go to the library where you can go back ten years or more. Even a 10 year old number is likely to be better than any guess you might make. The reason you want to do this is that the abstract does NOT just contain the same tables updated year after year but also contains a wealth of information mined from one-time studies that might only appear in one issue.

Gale Research which is now part of The Gale Group puts out a wide variety of reference publications including:

Associations Unlimited (for the trade association that hold trade shows in your product's industry or to find the people with needed expertise [in paper form this is known as *Encyclopedia of Associations*]),

Brands and Their Companies (for searching registered and non-registered, common law, trademarks),

Gale Business Resources (comprehensive histories and chronologies of major firms; in-depth views of 1,115 major industries — essays, rankings, market shares, associations and statistics),

Gale Database of Publications and Broadcast Media (descriptive and contact information for more than 63,000 newspapers, periodicals, newsletters, directories, TV and radio stations, and cable companies in the U.S. and Canada),

Gale Directory of Databases (guide to the electronic database industry with details on more than 13,000 databases accessible online, on CD-ROM, diskette or other media), and

Market Share Reporter (An annual compilation of reported market share data on companies, products, services.)

Talk to your local librarian about using paper or online versions of these Gale Group resources at the library. You can also try their web sites at www.gale.com or www.galegroup.com but after more than an hour of rummaging around I found no way at their web site to become (or even find out how to become) a subscriber to their databases. But if I were already a subscriber, and presumably had agreed to pay, I could get into the above resources online through their site. You can access their online publications through Dialog at www.dialog.com provided, of course, you sign up with Dialog and agree to pay—which you can do online there.

If you find an association relevant to your type product in the *Encyclopedia of Associations* you will probably be able to get some idea of the industry size by the number of members. You may also have found a good source of industry statistics. It won't hurt to ask. Many associations compile industry statistics for their members and publish the results. The catch may be that they only provide the information to members. Explain what you are doing and what you need the numbers for. If it seems plausible to whoever you talk to at the association headquarters that they can help you and the industry by giving you the numbers, you just might get them for free.

The *Market Share Reporter* is very similar to *Statistical Abstracts* but topically and data wise quite different. When you look at it, look up your product under its own name as well as "Retail-<appropriate category>" and Wholesale Trade-<appropriate category>" in order to be thorough. Each topic does not occur each year so look back each year for ten years or so. I found that in 1997 48% of jewelry sales occurred at jewelry stores and none of the 12 other seller categories had over 8% of the market.

Mediamark Research Inc. (MRI) and Simmons Market Research Bureau are the primary sources of demographic information regarding household consumption patterns for both products and media (TV, magazines, radio, etc.) in the U.S. Many large libraries will have these and they are almost certain to be found at community colleges and universities where business is taught. You should be very careful using these books. Follow the instructions in the books and look very closely at the headings on the page you are looking at.

In particular be certain that you understand whether numbers are dollar or volume numbers, what their magnitude is, and what the time period they represent is. Also be aware that many numbers are flagged as suspect (with * or **), which means the report authors don't trust them and that you should use them very conservatively when that is the only number available. For major magazines only, and for products, they break down readership and purchasing on almost any demographic characteristic you would want including marital status, age, earnings, job, education, years at current residence, etc.

Mediamark Research, Inc. has a web site at www.mediamark.com. Unfortunately my repeated attempts to register or locate anything online at their site have failed so I go to the nearby college campus and use the books for free. Their books are a 20 volume set with a very wide variety of products. If you have a cross-category product you may have to look in several books to find it.

MRI begins many categories with further product breakdowns than are used in their full pages of numbers. The full pages, for example, for jewelry are broken down only to the level of "Fine Jewelry" and "Costume Jewelry" but at the beginning of the category I can see that 16 million Fine Jewelry necklaces sold last year as did 14 million Costume Jewelry necklaces.

Demographically I find that Fine Jewelry (and presumably necklaces) are bought pretty evenly across the board, with households having annual income >$75,000 only 21% more likely than average households to purchase Fine Jewelry. In fact the only two demographic categories more than 20% less likely than average to buy Fine Jewelry are over 65 years old or have household incomes <$10,000.

If I look at Costume Jewelry I find that women are 56% more likely than the average "person" to buy Costume Jewelry and men are 61% less likely to. There are some men who buy Costume Jewelry for themselves (you can look it up) so that may mean that the rest just haven't learned yet, you don't buy costume jewelry for your wife (except by specific request or possibly strong hint).

Simmons Market Research Bureau has a web site at www.smrb.com but there is no online access to their data yet either. They at least do tell you the broad categories the products and brands they cover are in. Surprisingly, they don't even say how to get their reports or how much they cost so your best bet is to see them free at a good library. I think their breakdowns by media are a bit more complete and they may have more products but those are just impressions. They do have many more but often skinnier volumes than MRI.

Competitive Media Reporting (formerly LNA/Arbitron and often still referred to as LNA) have a site at www.cmr.com but it's not much more than a list of their products. If your product is a reasonable volume consumer good and you have major competitors this is the best place to find out how much the competitors are spending on advertising. (Hint: you'll probably need to do the same.) You must contact Competitive Media Reporting to get pricing, etc., information since it is not available online. Your best bet to find these reports is to go to a large academic library. The most useful volume to you will probably be *Ad $ Summary*.

For any of the above three resources your product, if it's truly new, probably won't have a listing. Also not all products or brands are covered. That does not mean these resources won't be useful to you. Think about the important characteristics of your users and of your product and see what other products with applicable characteristics might give you some insights to use patterns of your product by your kind of users.

Two good online listings of additional secondary marketing research (primary marketing research is done by direct contact with or observation of consumers, secondary through published resources) are the "Secondary Data Sources for Marketing Research" page at Vanderbilt University (www2000.ogsm.vanderbilt.edu/guide.html) and the "Marketing Research" page at Rutgers University (newark.rutgers.edu/~au/njseven.htm).

If you want a good academic overview of marketing you can go through the "Marketing 120, Principles of Marketing" course online at webster. gtcc.cc.nc.us/vcampus/mkt120/index.htm—just don't contact the professor or head to campus to take the test. The textbook, *Essentials of Marketing*, E. McCarthy and W. Perreault, Jr., 7th edition, provides a high-level, academic overview of marketing. There is not much "how to" in this course although the student is expected to complete a business plan during the semester.

Media Directories.

Many difficult markets can be reached by renting lists of relevant magazine readers or just advertising in the magazine. To find specialized magazines visit www.mediafinder.com. Mediafinder lets you search for key words or from a category list and locate magazines that would be of interest to your prospective buyers. Unfortunately, all you will get without becoming a subscriber is the publication name and possibly the subscription price. To get full info including searches for mailing lists or by publication type, subject category, target audience, circulation, advertising and list rental rates, frequency of issue, etc., you must subscribe at the current rate of $395 for 3 months or $995 for a year. Don't let the price tempt you to believe you shouldn't use the free search capability. Use the easy electronic search capabilities and write down the names of the magazines you find.

www.srds.com SRDS probably sets the advertising industry standard for coverage of various advertising media. At their site you can find out about their various publications and access online databases—but not for free. The SRDS (used to stand for Standard Rate and Data Service) materials are generally accessible by subscription. Their *Consumer Magazine Advertising Source* directory, for example, is $587 per year.

For complete sets of subscriptions you would spend several thousand dollars a year. You can also buy single copies of the directories but the typical prices are 50% or more of the annual subscription rate. Your best bet for looking at these is often to go to your nearest large library. If it has a business section these directories will often be found there. College business libraries virtually always have them.

Even small marketing offices (like mine) often have a set of their books so you might ask in the offices of a marketer you are seriously considering using. For a business type product the *Business Publication Advertising Source* directory is a good one.

Ulrich's International Periodicals Directory (electronically also known as Bowker's International Serials Database), and its brother and sister publications, are good sources to locate magazines of interest to your target market but the reporting on circulation numbers is far worse than SRDS. I suggest you start searching electronically on the Internet for these with the search string "ulrich periodical" to see what you come up with. With a little luck you may be able to get into the database on the Internet through a library that you are an authorized user of. These publications are also available in most major libraries.

www.publist.com This database actually lets you access everything it has online for free. I was able to find out, for example, that "American Jewelry Manufacturer" has a circulation of 5,600. Breadth wise I don't really know how their data stacks up compared to Ulrich's (they claim Ulrich's as one of their sources) or Mediafinder but I do know that not all listed publications have circulation figures. They do have contact information for the listed publications so you should always be able to contact the publication and ask for their circulation figures.

If the above list (or what you find elsewhere in this document) doesn't satisfy your need for secondary market research then I suggest you locate a copy of *Marketing Know-How, Your Guide to the Best Marketing Tools and Sources* by Peter Forancese (1996) (ISBN 0-936889-38-1). His list is overwhelming.

Primary Research

The three forms of primary research previously mentioned are focus groups, mall interventions, and test marketing. For more information on those check the index and go back to those sections. There are two other sort-of primary research methods that you should probably be aware of. These work best if your product is easily categorized and is not a dramatic departure from the norm.

The first is to ask store buyers, i.e., the people working in the store that are responsible for buying products to sell in the store. These people will usually provide their best answers after you have real products in real packaging, but some might be willing to talk to you at the model stage when you are most in need of a hint that all money invested in the invention won't be wasted.

Show them your model and ask their frank opinion of whether it would fit in their product line and whether it is appropriate to the demographics of their clientele. Also ask what volume they might expect it to sell at and the selling volumes of competing or at least possibly comparable products. Get the info in dollars and units if you can but don't be a pain because you can guess approximate conversions later by looking at the current store prices.

Before you leave, if possible, get the store's overall volume in dollars (you can probably look it up later if they don't give it to you). DON'T TAKE UP ANY MORE OF THEIR TIME THAN THAT unless they are obviously willingly volunteering it. Also, WITHOUT THEIR CONSENT, DO NOT argue with them about your invention. You will impress them much more favorably by genuinely thanking them for their input.

With the input of a few such store buyers and with industry numbers you can make some decent guesses about your possible annual sales volume after sales growth has ramped up to "normal." Keep in mind the issues of whether your invention is seasonal or regional when looking at industry numbers to make projections. A Christmas version of a butter dish is extremely unlikely to ramp up to the level of an every-day butter dish. A snowmobile gizmo won't sell well in Florida. And an Easter Bean Pot may only sell in Boston.

If the store buyer's don't like your invention you will have to decide how much you trust their judgement. If they are shooting you down due to competing functional products, then I would trust them a lot. If your invention is quite out of the norm, with only non-specific competition, I wouldn't trust them overly much. Remember that the Slinky was rejected by every original store buyer that was given the opportunity to sell it. There is another straw, grab it at your risk.

The second additional form of primary research you might do is to talk to multi-line manufacturers' reps or distributors. Ask questions similar to the ones above. You won't be able to get readily comparable dollar sales for the distributor from which you can easily match industry numbers, however, because each retailer that deals with the distributor is likely to deal with multiple distributors. Again, you are not there to waste their time or argue with them. Be considerate because you hope to be back to get an order from or through them later.

I believe that the above two approaches work best when your invention is 1) obviously superior in function (and usually cost) and 2) of significance to the store's or rep's clientele. An insignificant example might be a new pegboard hook while a significant example might be a replacement for soldering copper pipe joints. You might get a "ho-hum" on the pegboard hook even if they like it but they may promptly usher you to the door of an appropriate manufacturer for the non-soldering invention. (Note to quibblers: I don't know what the replacement to soldering is but for purposes of illustration it is superior to soldering on every imaginable criteria.)

Inside Secrets

Be very careful what you do with the information in the following paragraphs. In the wrong hands it could be dangerous. Sometimes it is not possible to make a profit on a product—by itself. Also sometimes it is not possible to make a profit on a customer—the first time they buy from you. How do you avoid going broke?

In the first case your best choice might be to offer a package deal where your product is included with others (of your or someone else's making). The trick here is to offer the package at a price that at least appears to be a significant bargain. The other products must also be such that they are not perceived to be valueless because the prospective buyer is likely to already have plenty of them. The most common "package" deal is probably the product plus a "free" book or two. Nowadays you also often see a product plus a "free" video tape. When produced in quantity, book(s) or videotape(s) add little to the cost of the package but they dramatically raise the perceived value. If your product uses a consumable that is commonly available you can also include a bunch of it (preferably a high quality brand of it) with your product.

If your product's consumable is your own consumable, your consumable probably should be your (hopefully protected) profit center, rather than your product being your profit center. If so, you should be able to give away a bunch of the consumable for free. Everyone probably knows the Gillette safety razor story: Gillette almost literally gave away the expensive razor and sold the cheap, but very profitable, blades. Make sure your test marketing is of the complete package, and make absolutely certain, if repeat consumable sales are your profit center, that your test marketing is big enough and long enough to verify the repeat sales. A 6-month test market is not unusual for large corporations when they know that repeat sales are crucial to success and they believe that the long test market will not give competitors any significant opportunities.

What if just getting a customer is always so expensive you lose money when a new customer buys? This is often more of a problem for retailers than product creators but it can be applicable. The "collectible" market is the classic example of this phenomenon. How does Franklin Mint or The Bradford Exchange make money if their advertising costs more than the money collected for the products they sell? Once they know you are a "collector" they will continually hit you with direct mail advertising for additional "collectibles" (which will also rarely ever be worth anything). If they only get you to buy with every tenth offer it still only costs them about $4.00 for that effort and you pay $29.95, or whatever, for the collectible (plus 3.99 shipping & handling). The direct costs for a plate and packaging, for example, are probably well under $3.00 so they still have a contribution margin of about $26.94 to cover their "all other expenses" including that money losing ad that got you interested. This concept is called "making money on the back-end."

Now let me tell you a really deep, dark secret. The real application of the above concept is not by inventors—it's by people (like me) who assist inventors. Not that all people who assist inventors work this way, but most do. First a low cost (e.g., $19.95) or even FREE offer (sometimes called bait) is dangled in front of you. Let's say you bite. Next you get the requested information and included with it are some additional offers and encouragement for you to take the next step.

It wouldn't even be unheard of for you to get a follow-up phone call encouraging you to take the next step. Not too big, not too small, but enough that you have a financial commitment. After that the next step is even bigger and you're encouraged even more. And you get the picture. This includes "ethical" patent attorneys and sly patent <u>agents</u> and marketers and even a fair number of thieves often disguised as "inventor clubs."

Keep in mind, a small ad in, say, Popular Science costs over $300 per month. Do you know any legitimate non-profit clubs that can afford that? It is not hard to set up a non-profit corporation in which the director is well paid enough to soak up anything that might have been corporate profit. Go ahead, get their free stuff but get really curious when they call or send further mailings that start suggesting you start sending them money for ANY kind of help.

Psychographic Categories

The number of inventor wannabes is growing and is probably mainly fueled by two distinct groups. First, let me say I know you don't see yourself in terms of a stereotype—despite the fact that you often see others in terms of stereotypes. Certainly not every inventor belongs to these groups but when I explain parts of their psychographic (new marketing buzzword) profiles **you should immediately see why the number of blatant rip-offs of inventors is increasing**. (Psychographics, which is really "Lifestyle Analysis" under a new name, is a categorization of market segmentation based on the activities, interests, and opinions that people express in their daily living.)

The first group is the retired or soon-to-be retired. This is historically the group that represents the most non-profitable patents. This group worked all their lives and went through World War II as the backbone of America. Most relied on Social Security and big company pensions to provide for them after retirement. The lucky ones have set aside $100,000 or more to help them "enjoy" retirement but these are few. The majority have some equity in their homes and less than $30,000 in the bank or investments.

Many are now realizing that the "reward" of pensions and Social Security is stacked against them due to expected inflation over the time frame they expect to live. "Saving Private Ryan" notwithstanding, they are also realizing that "recognition" is going to amount to a long forgotten ticker tape parade and a gold watch or less. An invention that "hit" it big would certainly turn things around.

The world has changed fast in these people's lifetimes: coal-oil to electricity, horse-and-wagon to space travel, mumblety-peg to Nintendo, handwriting to e-mail. For the most part, however, they have not led the way but have been led along by government and business. They have been helped by people who "know the ropes" before and when they get into "inventing" they discover real quick that facing the bureaucracy of the USPTO or gaining entry to corporate decision making are daunting.

The solution, of course, is to work with someone who does "know the ropes." If you have an invention idea and start wondering how to find someone who knows the ropes they practically fall into your lap. All you have to do is call an 800 number that you suddenly notice on the radio or TV. So you call. They immediately tell you NOT to tell them about your invention. They have some "confidentiality" forms you can fill out that will protect you and they will send you those with some free information on what they can do for you. You start to trust them.

You get the package and sure enough the forms are there providing iron-clad protection for your idea. And what's more, for FREE (or maybe a measly $29.95 processing fee) they will evaluate your idea when you return the form. Shortly you get back a glowing evaluation—but they make it absolutely clear that nobody can always successfully predict the marketplace. They obviously immediately recognized the value of your idea, and you know there are no guarantees—these guys know what they're doing! They'll share the investment costs with you. You do believe in your idea enough to invest in it don't you?

BANG! Your pride in your inventive genius and your investment savvy (not, of course, spelled "g-r-e-e-d"), force you to kill off that last nagging doubt and commit to the "reasonable" request of the guys that "know the ropes." Marketers actually have a name for folks (primarily men) who fit this psychographic profile. It is called "Grumpy Old Men" and it predates the movies of the same name. Typical loss is under $10,000 before quitting.

James E. White

The other category is the "Type A" personality. These are the high stress, heart attack prone, hard charging, predominantly male, employees that are always getting things done for other people. It is slowly dawning on them that they haven't done much for themselves. Social Security will NOT maintain their lifestyle. Their mutual fund company retirement plan investments have consistently underperformed the market for the last 10 years. They need a hit soon but they don't have the time.

The ownership of intellectual property seems the way to go. They don't have time to research and write a non-fiction book and fiction books require tremendous talent and it's a crap shoot anyway. A patentable idea where they can pay someone else to do the detail work seems like just the ticket. Typical loss is around $50,000 before quitting.

Just for completeness I'll mention two other groups (excluding corporate inventors) but these groups rarely lose money to hucksters. The first is the "idealist." Also predominantly male and mostly under 35, they see "right" quite clearly but have no money, often few friends, and have the intellect to get everything done "right" themselves. They buy books on patenting and write the application and process it through themselves. Then they wait for the world to see how "right" they are. It rarely happens. The smart ones these days also set up a web site to sell their patent. Takers are VERY rare.

The last group is the "entrepreneur," smart, personable, charming, going places, make-it-happen kind of people. These are also currently predominantly male but that is rapidly changing to a more even balance. They may invest some of their own money but the bulk of the money comes from people they have persuaded to invest in their dream. They believe in being in charge, often to the nth detail, and paying someone they completely control to do the job. Often they are not afraid to pay themselves a salary out of their investors' money. In the end, all most of them lose (of their own) is face—but it's only a temporary loss.

I know, you want to be thought of as Edison. But while people often idolize Edison's inventiveness they also instinctively mistrust "inventors." That instinctive mistrust is NOT misplaced because the record clearly shows that Edison was frequently a FRAUD as well as a successful inventor. Some inventors (AKA "snake oil salesmen") are only frauds. Evaluate yourself critically and honestly—just like you do your inventor friends. And don't think it's just me. Thomas Mosley, in *Marketing Your Invention*, devotes a whole chapter to the following 10 categories:

Psychographic Inventor Types (Incomplete)
1. The Paranoid Inventor
2. The Omnipotent Inventor
3. The Greedy Inventor
4. The Impatient Inventor
5. The Empty-Nest Inventor
6. The Emotional Inventor
7. The Deaf Inventor
8. The Flake Inventor
9. The Procrastinating Inventor
10. The Inventor Who Fears Success

The point is not to categorize you or for you to categorize yourself. The point is to recognize how you MIGHT appear to someone out to get your money and to let you see what your personal psychographic triggers are. Knowing that, you can take the steps necessary to protect yourself if someone should choose to attempt to trip your triggers to exploit you. (Heh heh heh!) You can also take the steps necessary to rid yourself of any blinders you've put on that would have kept you from creating a winning product or blocked you from succeeding with a winning product.

The Rules of the Game

As was mentioned earlier, it is very unlikely that you will change the industry's informal rules. It is also unlikely that you will change the legal rules of the industry. It's not that you can't, but to do so would often take far more time and expense than necessary to either get your invention made and distributed by following the rules or by devising your own mechanism and going around the rules.

Be aware that many things presented to you as "rules you MUST follow" are simply there to raise some barriers to the entry of new competition in a particular industry. But many "rules" are there to keep the industry's costs down so it can successfully compete with other alternatives. You will have to make some judgements as to what the real rules are and what motivates them. If you find a way around rules that are there to reduce competition, your competitors will hate you but your customers and the government will love you. How do you do this?

Just a couple of ideas, taken from *Millions from the Mind*, to get you thinking. The first is Wilkinson Sword and their new better stainless steel shaving blade. Rather than try to unseat Gillette from its dominant position in drug and grocery stores they initially sold their superior stainless steel blades in garden shops. Garden shops were where their blade superiority was well known in shears and clippers. It worked and in less than a year the consumer demand forced drug and grocery stores, the industry's normal distribution channel, to carry the Wilkinson Sword razor blades. Of course Gillette "discovered" stainless steel blades within 2 years.

Another example is the Interplak Electric Toothbrush. This is a superior electric toothbrush that was priced at $30 above its nearest competitor which the public mainly considered a gadget with minimal real benefits. Existing electric toothbrushes amused junior and maybe got him to brush a little better for a while but...so what. The Interplak people first got real scientific studies done AND "PEER REVIEW" PUBLISHED. Then they approached dentists with the published results and offered them the opportunity to sell the toothbrush directly or to take it to a drug store and let the drug store know that they would be recommending the toothbrush.

They also did a good job on the free PR and got some television and print media exposure due to the "proven" to save your teeth angle. The first year they lost $1 million on $4.5 million of sales but the loss would have been far greater, and probably the sales less, if they had nationwide advertising expenses. In the fourth year with profits in excess of $20 million they sold the company for $133 million cash. The product is now in the normal industry distribution channel.

So don't get mad at the industry rules—beat them. The first thing NOT to do is whine to an alternate channel "The X industry hates me, they're a bunch of !#$!%$$ jerks, can you help me?" First you must <u>identify a channel that is not only in your best interests but is in their best interests</u>. In the Wilkinson Sword case their blade's cash register counter display took virtually no space, provided a high demand "incidental" that increased the average sale per customer visit, and increased customer traffic. Of course the last two tapered off when the normal channel started carrying the blades. An "incidental," by the way, is something that you "incidentally" discover you want or remember you need after you have already gone to the store for something else. Batteries, candy bars, celebrity "news" rags, etc., that you see near cash registers almost everywhere are marketing "incidentals."

In the case of Interplak the dentist channel (which eventually included 15,000 of 65,000 dentists) gave the dentists extra profit, the drugstore channel (via dentist assurances of recommendations) got a pretty certain selling product (no guarantees), and the customer got assurance that the high price provided real value (due to dentist recommendations and real, published research). How do you work one of these deals yourself. YOU ABSO-LUTELY MUST approach it PROFESSIONALLY. If you can't do it yourself, hire someone who can or get an outside marketing firm (mine will do, ahem) to do the work.

A Couple of Hints

First **"follow the money"**—to it's source. I would estimate that 90% of inventors approach the people who are the end <u>receivers</u> of the current flow of money. In other words the people who WILL BE HURT THE WORST by the new invention. Your chances will improve dramatically if you approach the people who provide the money in the first place or people who get to keep a share along the way but are squeezed more by the *status quo* than they would be by your opportunity.

Second, provide documentation, preferably from independent and reliable sources, that makes it clear why your product is more in the interests of the people you approach, and their customer's, than current products.

Hopefully, you will ask the questions **"Who benefits from my invention the most?"** or **"Who <u>has the money</u> and will <u>substantially gain</u> from my invention?"** when figuring out what your initial marketing channel should be. That way you will be able to get the most bang for your marketing buck when you are starting out and need that bang the most. If you plan on licensing rather than subcontract-manufacturing the best questions might be "Who is number 2 or 3 in the industry?" or "Who is hungry and receptive to ideas that will help take them to number 1?"

James E. White

CHAPTER 9
Advertising Claims for Your Invention

No Lies Please

The short rule regarding the claims you make for your invention is: DON'T MAKE ANY CLAIM YOU CANNOT SUBSTANTIATE VIA EXISTING DOCUMENTATION OR AN OBJECTIVE TEST. That is pretty simple—it means don't lie. But, you say, "All marketers are liars." That is almost universally false. A small subset of marketers often do provide you with information which you fool yourself with but that is all. (Well, okay, there are some marketers that are outright liars.) The major areas that people fool themselves in are greed, love/sex, and vanity.

I manufacture a 1 cent pill that can help you lose weight and get the attention of the lover of your dreams. Distributorships are available but going so fast they are no longer available in some counties. For high profit resale information just check the box when you send $19.95 for your sample bottle of 200 fully guaranteed pills.

Note that was not really a commercial interruption, please do not send $19.95, my hypothetical pill does not exist. But you should see that the writing does at least attempt to strongly appeal to greed and love/sex and possibly vanity. It does not, however, promise anything except to deliver 200 "can help, fully guaranteed" 1 cent pills for about 10 cents each and, at a minimum, information on how to become a distributor. Any belief that anything else was promised was generated entirely in your head.

Ask yourself this: "Do I believe every claim I hear?" Of course you don't, because you've heard some doozies over the years. You probably often ignore factual claims because you know the claimant has an "obvious" agenda in their own favor that the claim supports. Everybody's got an agenda—like, you want me to believe, you don't?

You Gotta Know Your Limitations—The FTC

The Federal Trade Commission (FTC) (www.ftc.gov) is constantly on the lookout for consumer fraud—i.e., advertisers selling but not delivering what they promise. Unfortunately, most FTC "watchdog" action is taken only in response to a pattern of complaints. A number of advertisers do try to skate as close to the thin edge of misrepresentation as possible but, if you take the time to look at the next 50 ads, commercials, etc. you come across you will probably find that most companies play the game pretty safely.

The reason is simple—successful companies want your business again and again and again and they want you to recommend them to your friends. Why? Because the solidest foundation for a wealthy businessman is repeat and referral customers. First time and one-time customers are often gotten at a loss. True rip-off artists that stay within the law have very tough going because they must be constantly inventing new rip-offs that (just like inventions) may or may not pay off. Maybe we should call those semi-rip-off artists. The blatant fraud rip-off artists on a mass consumer scale are usually "one-shot" people that get caught during their second or third try at fraud (usually because they got just a little greedier).

The FTC web site is a very good place to go look at the rules and regulations governing advertising and to get information on the latest scams that are being perpetrated.

Why mention all this? Because, if you want to get rich from your invention, you must abide by the law and plan on gaining your wealth over an extended (say 3 years plus) period of time. The FTC catching you in a lie is the least of your worries—your biggest worry should be getting enough satisfied customers that you can make a profit.

You also need to be aware that the FTC (and the public usually) often consider some types of claims as irrelevant. These types of claims are "subjective" in nature rather than "objective." The most typical of the subjective (i.e., in the eye of the beholder) claims is "best." Usually the public and the FTC will give a manufacturer the latitude to claim their product is "best" as long as it is nearly equal to competitive products or has at least one significant attribute that may arguably be better than nearly all of the competitive products or if, in a fair comparison test, more consumers/judges/whatever rated it best on some criteria.

James E. White

On the other hand, advertising that over-hypes a product using mostly subjective criteria is usually NOT successful—"*Caveat emptor*" and "If it looks too good to be true, it probably is" being well known sayings these days.

Please don't just believe that all you have to do is give a demonstration to someone and they will believe you. I had a farmer friend who was in the market for a new tractor. He was talking to a salesman about various features and the salesman brought up the (extra cost?) easy-change tire system and started to give a demo right there in the showroom. My friend's comment was (approximately) "Would you mind if I asked you to hold off on that demonstration till I have a flat tire in a low spot in my field in April when there is a light drizzle, its 35 degrees out, and the wind is gusting to 30 miles an hour."

Aside from the fact that you are likely to do the demo in ideal (rather than real-world) conditions, people are also aware how easy it is to perform unseen "magic" when the inventor is in total control. A thermometer good for 400 degrees has its scale replaced with one reading to 2500 degrees or the "perpetual motion" machine has a little man inside for example. Of course that would be fraud—but overzealousness has produced even worse crimes. Have you ever fooled anyone to get them to believe you? If you have inventor friends, do you always believe them?

Objective Claims

Where you should really concentrate your effort is on the objective claims you can make. These are claims that can be tested and proved to be true or false. Only make claims that you know can be proven true. The FTC can require you to provide documentation that verifies the claim. The FTC will not accept documentation that simply asserts that the claim is (or is believed to be) true. While it is not mandatory, I strongly recommend that you spend the money necessary for an impartial (except that they know who is paying for it) entity (testing lab, consumer surveyor, etc.) to validate your claims.

To have really strong claims you can also have a third party submit your and your competitors products for simultaneous independent testing. If you do this however, the FTC requires that the competitive products be reasonably comparable major brands— Brand X does not qualify if better ones are commonly available. If your invention is a medical device, the federal Food and Drug Administration (and the law) mandate that your claims be validated by rigorous scientific testing AND THAT ANY POSSIBLE SIDE EFFECTS BE DISCLOSED. Anecdotal evidence (I did it and it worked for me) is not acceptable.

The FDA web site (www.fda.gov) is excellent and very extensive but it will take some work to dig out what you want because there is no consistency across the pages prepared by the various branches of the FDA. In one branch the page you want may be found under "Regulatory Guidance" while in another it is found under "Laws or Regulations Applicable to or Administered by..."

But, you say, what about the powers of copper bracelets or magnetic shoes? Read their ads closely. You'll see "Mrs. X says," or "believed to **stop aging permanently**" or whatever. If the products appear to do no harm and ads really don't make any miracle promises there is virtually nothing the FTC can do. The product is usually a "fad" product with a profitable maximum life of about 2 years. (Or at least it will take about that for the copycats to beat the profit margins down to near 0.)

I have no proof but I'm fairly certain the following psychological mumbo-jumbo is applicable. First the buyers are typically the curious (with money), the (near) desperate, and the pseudo-desperate. The curious will try the product with no effect but they won't have the patience to continue using it for the 3 months or whatever the manufacturer suggests is required as the minimum. It winds up in the junk drawer, and since the money was incidental and the curiosity was sort of satisfied, there are no complaints.

The desperate need real relief and will try anything. It's different so at first the product seems like it might be doing something. In reality the placebo effect is working its magic but in the long run it can't sustain itself. The product winds up in the junk drawer but the gamble was worth the price so no complaints.

The pseudo-desperate get "instant" relief which they vociferously tell anybody and everybody about in the same way they let everybody know about their "pain" before. The success (or failure) of the product is not the point—people paying attention to them is. The "newness impact" of the product wears off so a new and "better" (after all we can't admit failure or a waste of money for the product) "cure" is sought to recapture the waning attention. The marketers collect the testimonials of a few of these people shortly after they get the product in order to feed their next marketing campaigns. Again to the junk drawer with no complaints.

If that is your kind of invention please don't call me because I will require bona fide scientific testing before I'll help sell the product.

As an exercise, can you argue that the above type product IS in the best interests of the consumer? Do that now before you read my "justifications" in the next sentence. To the curious it was an amusement, to the desperate it was a ray of hope to get them through another day and a worthwhile gamble, and to the pseudo-desperate it was an attention getter. Were they wrong in dumping money into the pockets of the inventor/marketer? Am I a fool for not accepting those kinds of products?

Fuzzy Qualifiers

You should not try to make a claim and surround it with fuzzy qualifiers. The fuzzy qualifiers may actually have the effect of increasing your testing cost should you be called on to do it. For example "cleans a grill in 3 minutes" is a cheap test to perform; 1 grill (with presumably unbiased selection) and 3 minutes of time will give you a proved/not proved answer. If it takes 3 minutes and 10 seconds your claim is obviously false but may still be considered within acceptable "hype" tolerances.

On the other hand the claim "you'll typically clean your grill in less than 3 minutes" can require considerably more effort to prove. Now you probably need to get a representative set of grills (big, small, fairly clean, very dirty, etc.) and time the cleaning of each of them with everyday users rather than your factory experts. An average time of 3 minutes and 10 seconds may not be considered tolerable for this claim.

However, if qualifiers are essential for the claim to have any validity (or even believability) you MUST include them with the claim—at least to avoid being ignored. For example "my invention will take anything from 0 degrees to 3,000 degrees in 3 seconds" is a pretty bold claim. Nobody will believe it and you clearly cannot prove it true. If I asked you to prove it by raising the temperature of the moon to 3,000 degrees Celsius I am certain you could not do it. I would be willing to bet you $1,000 you couldn't even do it for a 2-inch cube of steel.

Without putting qualifiers on such a claim you pretty much guarantee that you will lose 99.999% of all potential sales because you will simply be tuned out. If you know what you can substantiate—"I have invented a heater that outputs .01 BTU per second at 3,000 degrees Fahrenheit within 3 seconds of startup"—then state it clearly and a lot more people will be interested. **Remember, as the inventor it is in your best interests to get your message across. It is not your target's responsibility to correctly understand you.**

Owning the Patent Is Enough?

Many inventors apparently believe that simply owning the patent for an "invention" with many bold patent claims (which they may actually believe) will entice someone to license the "invention" and presumably pay the inventor money. This is simply not true. A quick search of the Internet will turn up at least 5 sites by patent attorneys that offer their clients' "inventions" for licensing. Some of the "inventions" actually work but there are a large number that no one of any intelligence (possibly except the inventor) will believe will work.

Looking at the device with a little common sense (which is occasionally known to be wrong) and/or real life experience will lead most observers to believe that the "invention" will not <u>satisfactorily</u> solve the problem it claims to solve or that it simply won't work. Too many patents make blatantly unsubstantiatable claims and are for un-developed and untested "inventions." In clear English, the inventor wasted their money and/or time getting the (worthless) patent.

Remember, as discussed in Chapter 2, the USPTO errs on the side of granting the patent if no "prior art" is found and the inventor asserts that it works or often just that they believe it works (except for perpetual motion). **Just because the USPTO granted the patent with the (false) claims that does NOT validate them in the eyes of the FTC** (or the USPTO or any other government body or the consuming public in the long run).

Just to repeat, DO NOT EVER put "Patented" on your product (unless that is the form of notice for some foreign country where you HAVE A PATENT on the invention and that is where you are selling it). In the U.S., "Patented" is not an acceptable form of notice. It is clearly illegal if you don't have a U.S. patent for the product. DO NOT claim "Patent Pending" if you have not submitted a Provisional or a full Patent Application to the USPTO. The court will not be amused if you are challenged on such a claim and you say "It's pending in Abuwanda."

James E. White

I always check, right at my desk, any inventor who makes a claim to me that they have a patent on the idea they just told me. All unexpired U.S. patents can be searched on the Internet at www.uspto.gov and the database is updated weekly, usually at the same time the grant announcement is published. If you were going to market someone else's invention you would immediately check out their patent claims just like I do, right? If you are going to market something, you want to be sure you have a chance at the profits rather than letting them all go to a knockoff maker, right?

Yet I'm told all the time "I've patented that" and 80% of the time I find there is no patent. First, I believe the individual is a liar and I will discount his other claims also. Second, I believe that what such liars usually mean is "I plan someday maybe to probably make some kind of effort toward seeing what it might take to try to get a patent if it's not too expensive or inconvenient but in the meantime could you worship me anyway and call me a genius?" If you think that is harsh then "the shoe probably fits"; if you think that harangue is only fair and right then you probably have a bright future. Stick to the absolute truth where patents are concerned.

You should also be aware that, from a marketing perspective, your claims must be tied to the best interests of buyers. "Made of heat treated 330 C Steel" is a wasted claim since few people will know what benefits heat treated 330 C Steel has relative to any other steel (or material for that matter). "Making this product of heat treated 330 C Steel allows us to guarantee that it will hold its cutting edge for 20 years—even when used daily on the hardest granite." Now we know something about how heat treated 330 C Steel benefits us even if we don't know that 330 C Steel is just a designation that I made up.

Your Very Own Institute

While you can create your own club or organization or even something you designate as an "Institute," and then wear some hat in that institute while you make some proclamation or claim that you intend to advertise the hell out of, I don't recommend it. It is not illegal although I suspect you'll find a fair percentage of the population who believe it is. It is clearly in the broad gray area of ethics but approaching the darker end of the area.

If, when wearing your "Institute" hat you make a blatantly fraudulent claim you can still get into real trouble, but if your claim is fuzzy and your ads hype up the "Institute's" proclamation your likely "punishment," should the FTC ever get enough complaints to take any action, will probably be to include a suitable, small print, footnote in your future advertising noting the relationship between the company and the "Institute."

Worse is if customers lose faith in you and take their business elsewhere. Why risk it? **If your best claim for your product has to be vague why not work on another invention that really gives the customer something?** Unless you are pathologically immoral, you'll feel better about your profits too.

James E. White

CHAPTER 10

Getting Professional Help

If you do have the talent (natural or learned) to write sound patents, design working devices, finesse a consumer-pleasing industrial design, build models, machine prototypes, develop packaging, write ads, etc., by all means use it. If you don't have those talents, trying to pretend you do just to get by on the cheap may be the most expensive decision you ever make.

You Pay for It

Professional help is help that you <u>pay for</u> from an individual or firm with expertise <u>in the area you are paying for the help in</u>. Too often inventors seem to think that any professional will give great professional advice on anything. Usually they do give good professional advice in the area of their expertise and usually they try to make it clear that any other "advice" they give is just jawboning.

Since the "other" advice is free is it a good deal? The answer, of course, is maybe or maybe not. It may prove to be sound because it really is good advice or it may just be luck or it may be just plain wrong. For some professions (e.g., legal or engineering) in many states it is against the law to provide professional (paid) advice without state or other certification. If you don't know, find out what is appropriate for your state.

When you are working with your patent attorney, don't expect them to evaluate the engineering merit of your invention unless they are also engineers (and so licensed in your state if required) AND YOU PAY THEM FOR THAT SERVICE. While it is true that the USPTO mandates that attorneys registered to practice before them have undergraduate degrees in engineering or science or "prove" some "equivalent" experience, my understanding is that "equivalent proof" is handled somewhat loosely. If you were the USPTO would you want to be in the business of calling a lawyer a liar if they claimed technical expertise? I don't think engineers are taught "hingedly affixed" and other such patent phrases anyway.

Patent Attorneys or Patent Agents

Patent attorneys tell me that inventors are always asking for their opinion on how great the idea is, or if they think it will sell. Some will give a personal opinion and make it clear that is what it is, a personal opinion. Most however try to give a non-committal answer such as "You need to work with a marketer" or "That's not my field." They have a vested interest in not turning you away because they stand to make a buck preparing and following through on the patent for you regardless of whether you ever get (or even can get) your invention to where it sells enough to recover the patenting costs.

If you go out on the Internet and do a search you can find law firms and lawyers that will give you an opinion on the marketability of your invention—often after you've signed away a percentage of future profits should they decide to help you. They do the patent work and the financing and take a lot of the hassles out of it for you, but they get the biggest percent of the profits for their trouble and they reject 95%, or more, of all inventions offered to them because they only want to work on s that are reasonably certain to get them a profit.

Some of these attorneys and venture groups limit themselves to only certain types of inventions for which they have built up manufacturing and distribution channel contacts over the years. Of course, they will NOT tell you that the rejected inventions WON'T SELL, just that they don't want to handle them. Your rejected invention MIGHT sell; you'll have to proceed on what you think (and, of course, the steps in this book).

When working with an attorney, be aware that it is not illegal to patent an invention that does not work, and since you are in charge of your patent attorney, they can and will simply put your invention into proper legal form and take it through the patent process. If that is what you want, fine. <u>It is also not illegal for the patent office to grant you an invalid patent.</u> Whether your patent is invalid or not, it will be your, not the patent office's, responsibility to meet any challenges to your patent.

Your attorney, of course, will also want more fees (usually up-front) **IF** they agree to TRY to defeat the challenges to your patent. If you want to patent an invention that works, however, you may want to get some outside professionals (perhaps engineers) to assist you with development before you apply for the patent. File a "Disclosure Document" or get witnessed "read and understood" statements; then get non-disclosure agreements from the engineers if you want some protection.

Non-Disclosures

Just a word or two more about non-disclosure agreements and professionals. Not all professionals will readily sign them. That does not mean they are dishonest or thieves. <u>A thief will instantly and willingly put an illegible scrawl on any document you present. They know the law and that they can stay just enough inside it to avoid a criminal (at taxpayer expense) investigation and prosecution. They also know they can probably take you for enough money that you are unlikely to have the financial resources left to mount a (possibly failing) civil action against them.</u> (Attorneys and Patent Agents are obligated NOT to disclose your invention, by the way, so you don't need to ask them to sign a non-disclosure.)

What to do? Keep your bound inventor's notebook up to date with dates and times and who and where you met and what you discussed and anyone else who you were introduced to or met. Keep paper records of correspondence and log phone conversations you have when setting up appointments, etc. AND KEEP THE PHONE BILLS. In the unlikely event there is a problem you will have far more credibility with such documentation than you would in a "their word against yours" contest.

There are university evaluation programs that totally refuse to sign any kind of confidentiality agreement and I've encountered one estimator/drafter/prototype maker who refuses to sign any non-disclosure agreement except a commercially purchased one. To some extent I can't blame them. To do the job, a marketing analyst, for example, must disclose the invention to potential purchasers, and some attorney agreements are so convoluted you aren't sure if you attribute all your future ideas to the inventor you are signing the non-disclosure for or not. Regardless of how careful the person who signs your agreement is, they know you can still sue and claim injury. Win or lose, the hassle of a law suit is almost never worth it. Been there, done that.

Engineers

When working with an engineer or other technical professional be aware that the technology field is huge today. Nobody knows everything about everything. If you need materials or use technologies that are even slightly out of the ordinary you will have to spend some time locating the companies and individuals that can work with you. Most likely you will have to ask the people who don't have the expertise to refer you to someone who does or who might know someone who does. You will find all sorts of levels of cooperation, from:

"the manufacturer of the machines I use makes one that will do what you want and I don't have one, but for competitive reasons, we can't disclose who manufacturers our machines" to,

"my company makes machines that will do what you want but here's the phone number of the guy you want to talk to at XXX company. Their technology makes prototyping and development work much cheaper."

I've paraphrased, of course, but I've gotten both of those responses: the second from the owner and chief sales officer!

You will often find that after talking to the engineers at one or two manufacturers you either know more about the specifics of the materials you need, or at least appear to, than the next engineer you talk to does. This is not unusual: the people you talk to may, but often need not, have engineering degrees. I don't believe all states have laws defining who can be called an engineer and even in states that do you may be turned over to the "design" or "technical" staff and they don't discourage your assumption that they are engineers. They often just receive and pass on engineering specifications that other manufacturers' engineers have explicitly given them.

They expect you to be as knowledgeable as those engineers. This is particularly true if you are planning to shop out specific components or processes, such as springs or tempering, to firms specializing in those areas. I have occasionally gotten exactly opposite answers from "experts" at the same company so I can sympathize with you. If you are not careful, you may be talking to the sales representative who believes it is his responsibility to get your project (and money) in the door and the back room's responsibility to make the deal work. You'll have to use your judgement, do some independent study, and challenge the company to provide corroboration for their statements that don't make sense to you. Of course, you can always take your money and move on.

James E. White

University Experts

For some technologies you may get referrals to university experts who may or may not expect a consulting fee. Most won't charge for an hour or two of providing or directing you to information, but when it gets to serious research or other effort they will undoubtedly ask for funding, perhaps for a project from which they hope to publish the results. Evaluate what you want carefully and whether it might hurt you for the results to be published before striking a written agreement with them. (Published, proven results can work to your marketing advantage too.)

Also be aware that if they are to be using university equipment you will probably be required to contract with the university—and believe me, they will have hoops for you to jump through. The farther the expert's location is from yours, the less likely you will be able to ensure that you are getting what you want versus what they want. And, of course, just because they are a university professor does not mean they are honest, knowledgeable, or skilled. If you're not comfortable with who you are dealing with keep looking.

Large corporations get into university research agreements all the time and they generally have the resources to provide the funding and the clout to get the researcher to do the research they want. (If their research has an "agenda," such as proving material X does NOT cause cancer, they and their researcher are occasionally embarrassed when the desired result is obtained but later turns out to be false.)

If you don't have the resources, there are government grant programs available that may, with a lot of paperwork on your part, be willing to fund the research IF your researcher has the desired credentials. A couple of places to start researching for these are: National Technology Transfer Center at www.nttc.edu, Small Business Innovation Research Programs at www.sba.gov/sbir, and the Invention to Innovation Program of the Department of Energy at www.oit.doe.gov/inventions.

DO NOT pay $5,000 up-front to someone promising to try to get you a grant. Even paying $60 for a list of "sources" is probably a waste of money. Also visit your local library, Small Business Administration office (check www.sba.gov for locations), and/or science and technology university and start asking questions. No matter what, you will quickly find that grant money takes a lot of effort to get and that it can usually only be gotten in months and years, not the days or weeks some would have you believe.

In general, if your first invention or two require the expertise only university researchers can provide, I would recommend you skip those ideas for now and develop another one first. If you are not comfortable with senseless government bureaucrats and paperwork you probably shouldn't start down this path at all. With universities you also need to be very wary of any contracts that might be presented to you.

I would highly recommend that you get an attorney and that you negotiate as necessary. All contracts are negotiable: any university administrator who tries to convince you otherwise could perhaps technically be charged with fraud or you could later get the contract voided if a jury believed you were "coerced" into accepting the agreement. Most likely, the administrator is attempting to get the university some financial rights to at least a piece of your invention or ideas to put into their "Technology Transfer" portfolio.

Marketers and Licensing Agents

When working with a marketer or other business advisors (except accounting) be aware that most states do not regulate marketers or other business advisors. My advice here is for you to read over the "Contact Me" chapter in this book to get a feel for what you might be willing to do based on the kind of options I provide. Think clearly about what you want and find someone you believe you can trust to provide that.

You should also be aware that many marketers, myself among them, will NOT just blindly put any claims you make into ads or other marketing materials. We will insist on verifying them with our own eyes or having bona fide third-party verification.

Fixed Payment vs a Percent

If you want the bulk of the profits from your invention you will have to be prepared to pay, usually monthly or up-front, for services received. If you are willing to share the profits—perhaps even apportioning a lesser share to yourself—you may want to grant some form of partial ownership to your business consultants. **It is not unusual for start-up firms to get professional help in exchange for stock in the incorporated firm.** How

James E. White

much control are you willing to give up? If you don't retain more than 50% of the shares you may not have any control. (If you think about giving away percents of the company get an attorney familiar with securities law involved early in your thinking.)

If you want to retain control but reward marketing people that work with you, you may want to give them a (fat) commission. (Be aware that a fat commission <u>offer</u> might only get you 40 <u>total</u> hours of work on a product that won't sell.) **Fat commissions DO NOT guarantee sales, just initial interest from the marketer.** The bottom line here is that you are in most control when it is your money and effort on the line. The more other people have money and unpaid effort on the line, the more control (and profits) you should be willing to give up.

Whatever you do be very very cautions about demanding money up-front for yourself. To get that money you will likely have to give up part of your company or profits to the payee as an investor or you will be licensing rights at a lower long-term royalty than you would otherwise have to. When you demand money up-front a business person's natural tendency is to believe that YOU DO NOT BELIEVE in the long-term future profitability of your invention. The most likely result is that you don't get the help you need or the agreement you want doesn't happen.

CHAPTER 11

Getting Free(?) Help

Free help from other inventors is often plentiful. Should you accept it? I highly recommend simple written agreements—even among friends. True, an agreement may be embarrassing to ask for but it may ensure years of friendship and peaceful co-existence when the start-up responsibilities are over and it is time to share the wealth. You might be amazed how well some people can divide by 2 when they formerly couldn't do something as simple as trip over a rock. Think of the agreement as kind of like lending to an in-law—with a formal interest rate and payment schedule. **Fortunately(?), most inventors working together never successfully get a product to market so there are no profits to argue over.**

Good Intentions

Also keep in mind that the best intentioned help is not worth anything if it doesn't get the job done. At best you may get some positive results but the odds are they won't be near what you could get from someone who knew something about what they were doing. The more typical case, I suspect, is that the two (or more) of you work together for a while till your abilities are exceeded, then the project sort of slowly drifts into a back burner state from which it is likely to never recover.

If everybody is working day jobs, dealing with kids on evenings and weekends, and only working on the project "as time permits" the chances of expeditiously getting to the profits while enthusiasm is still high are slim. A paid (or otherwise compensated) individual, on the other hand, is likely to accomplish more in an 8 hour day than 2 volunteers can complete in 4 sessions of 4 hours each.

The three worst case scenarios are, 1) that you spend hours working with free people on projects which don't have any profit potential in the first place, 2) that you never complete a project that has tremendous profit potential, or 3) the project realizes its profit potential but it is all lost in litigation between "friends." If your available free expertise is in tinkering, that is what you should expect to get, lots of tinkering. If the free expertise is in law you will

James E. White

probably be incorporated before you know it and will have to devote time as required for the record keeping the government **requires** of business entities. You are likely to also get lots of fabulously crafted contracts which may have the unwanted side effect of scaring off people you want to collaborate with.

Agreement Issues

I recommend an agreement that includes at least the following points.

Split—Who gets what percent of the eventual profits from the invention? It is also a good idea to define how you will split the work. It won't hurt if you base your split partly on the basis of the labor each contributes, although I suggest that regardless of the labor levels there be a minimum share percent, but that the share percent require some minimum level of effort. For example, with 2 partners, regardless of who is the main idea person, the minimum for the one doing the least work might be 30% but that can drop 5% for each month they work less than 10 hours. There is nothing that prevents partners from adding addenda to the agreement that say that "March, when no work was needed, does not drop the minimum percent by 5%."

Careful records must be maintained by each partner (in your inventor's notebook is the ideal place) AND in logbooks at each work site. If you do the "lab" work in one partner's garage and the Internet "research" in another's basement you have two work sites. Ask any university research professor or business lab scientist and they will tell you this practice is normal if not mandated.

If you agree to allow independent work you will quickly see that a partnership becomes WORSE than a marriage from a responsibility and sharing perspective. What prevents one partner from checking into his garage for 5 hours of football watching? **If you don't trust your partner, don't sign up for a partnership.**

Death—How are the splits shifted if one or more partners dies? This is particularly a problem if the death occurs during development. It's your idea, you take on an expert partner, she dies after 10 minutes without contributing anything, now what? On this one you MUST consult an attorney IF the splits are going to be adjusted so that the share of the decedent DOES NOT go to the spouse (or possibly other heirs). Some states have laws that require the heir(s) to sign statements to acknowledge that they don't have any claim to the decedent's share.

It will be even worse if the death occurs while a divorce is in progress. As a starting point for negotiation try cutting the spouse back to 5% or, if the decedent was the main idea person, to 20%. Also schedule an annual renegotiation of the split adjustments that are to occur on death and also plan to renegotiate as product launch nears. How to handle stubborn (or non-) negotiators is covered later.

Remind all parties to update their wills (They do have wills, right?). Remember, the issue you are addressing is the value of intellectual property which can last long after any individual's death. Also keep in mind that your federal congress people (and some state congress people) have given the IRS (or other government bodies) the authority to sell off your intellectual property on death to satisfy their own desire for funds. **Thousands of businesses have been killed to satisfy politicians' lust for other people's money.**

Expenses and payback—Agree who will pay for expenses and how they will be paid back if you ever make a profit. My suggestion is that you agree that any expenses under $100 be paid by whoever happens to be handy and that individual decisions on expenses over $100 be done jointly. If all partners balance expenses (e.g., everyone antes up $10,000 into a common pool to start and equally contributes to refreshing it as needed) then I suggest profits simply be dumped back into the "pool" till it can be dissolved. If expenses must be significantly unbalanced then I suggest that contributions be repaid equally till the short contributors are paid back, then the long contributors be paid back $2.00 for every additional $1.00 they contributed before "profits" start piling up for the split percentages. If one partner is primarily the moneybags and the expenses are quite large (say $100,000) negotiating this kind of payback can make it possible for the idea person to get a larger percent of the eventual profits yet reward the real risk taker.

Patentable ideas—Each person who first contributes a functional idea that is patentable and is incorporated into the invention gets their name on the patent application as a co-inventor. Be forewarned, a person brought on as a "moneybags only" will often want to be a co-inventor on the patent. Have them talk to your patent attorney or one of their choice so that it can be pointed out to them that if they insist their name go on the application as co-inventor, their investment will be in a worthless patent because it can easily be declared invalid. Showing the moneybags person the oath is often enough. Intelligent investors will opt to accept the rewards and stay off the patent.

Be wary of who is at your "inventing" meetings and keep good notebooks. If you must be impolite and demand that people leave the room, do it. You'll probably regret it less than the alternative consequences. You can tactfully(?) point out to them that it is a purely business decision on your part. If they want to they can come up with their own invention and get their own partners and perhaps even beat you to success.

Co-inventor status can be tricky and I encourage you to discuss it relative to your invention with your attorney. YOUR OFFICIAL PARTNERSHIP AGREEMENT or this agreement we are now discussing should spell out OWNERSHIP of the invention/patent so that the issue of who the profits go to does not cloud people's brains. Co-inventor should be a mere technical issue based on functional contribution to claims, try to keep egos out of it.

Invention refinements and suggestions alone would generally not count unless the reduction to practice was obvious to everyone from the suggestion. For example, if the suggestion is that the lawn mower should store hanging on the wall, it is not obvious how to do that. If the suggestion is to use a flat instead of a round wire spring because it provides both the flexibility function and sufficient lateral stability to eliminate two stabilizing pins then you might very well be in co-inventor territory.

Co-inventor status is NOT dependent on whether you hired a person for engineering services. Your agreement with them should make clear your agreed payment to them constitutes their entire and only compensation and that they assign ownership of all improvements they make to you. If they are simply doing the "skilled in the art" engineering to execute your ideas that also does not constitute a reason for co-inventor status. If they are solving problems you hadn't anticipated and their solutions are "novel," "non-obvious," "functional" contributions to the invention and beyond "prior art" and "ordinary skill" you probably should recognize them as co-inventors on the patent.

Binding Arbitration

Arbitration—I would insist on a binding arbitration clause both for disagreements arising from the agreement and, if agreement can't otherwise be reached, for any negotiation or re-negotiation your agreement might call for. Binding arbitration is not free. You have to pay some money (last I checked it was under $200) to initiate the process then additional fees for presenting the evidence and issues and the decision. For a 2-3 hour process, and where the amount at stake is not approaching or over $100,000, costs should be well under $1,000 each unless participants have their own (at their own expense) attorneys. Arbitrators will probably be able to spot the "unreasonable" person within the first 10 minutes.

Per the rules, the arbitrator has the right to saddle the "unreasonable" party with all the arbitration expenses. However, to get that to happen you should probably put in a written request for that as part of the settlement; otherwise, to keep their own lives peaceful, the arbitrator will be more inclined to do a simple 50/50 split. The web site for the American Arbitration Association is www.adr.org and is well worth a look. Binding arbitration MAY NOT conclude with a result that you believe is "fair" but you'll have to live with it. **Again, the person with the best records is most likely to get "fair" treatment.**

When binding arbitration first got started it often had the reputation that the majority of the participating attorneys acting as arbitrators were failures that wouldn't otherwise have paying work. I think things have come a long way since then so that today it's highly probable that your arbitrator(s) will be sincerely dedicated to "justice" rather than their own welfare or the legal "nits" that often seem to govern judicial results these days. The "spirit of the law" is alive and well in arbitration.

One other caveat here, I recommend you go with the American Arbitration Association suggested contract clause (below) or a very near approximation of it. Their arbitrator selection rules are geared to help prevent any bias that an arbitrator might have in favor of a possible repeat "client" who they figure will appoint them again and again (thus generating a nice stream of fees) in exchange for favorable rulings. If you look at the contract you signed when you signed up with a securities brokerage house you may see that they explicitly write language into their binding arbitration clauses to ensure they get a favorable bias.

American Arbitration Association binding arbitration clause:

Any controversy or claim arising out of or relating to this contract, or the breach thereof, shall be settled by arbitration administered by the American Arbitration Association under its Commercial

Arbitration Rules, and judgment on the award rendered by the arbitrator(s) may be entered in any court having jurisdiction thereof.

You have already been cautioned about "FREE" idea evaluation services in the "Pitfalls to Avoid" section so go back and read it there if you skipped that section.

CHAPTER 12

Contacting James E. White & Assoc.

Several pre-publication reviewers of this book pointed out that it occasionally seemed that my frequently negative tone was also quite condescending. I have attempted to re-write so that it is not condescending but you may still feel I have failed. That is my fault. I think of you, and all inventors, as creative equals.

On the other hand, I don't want to deal with inventors who refuse to seriously try to answer the "Will it sell?" question before big development, patenting, and other time and dollar costs are incurred. I could deal with such inventors quite comfortably for the rest of my life with money up-front for marketing but there are plenty of marketers out there who will do that with no (or hollow) guarantees to the inventor and no risk to themselves.

My experience with creative people is that we too often reject vicarious learning from other people's mistakes in favor of doing it the hard way (i.e., making the mistakes ourselves). Maybe I'll learn the hard way that sugar attracts more flies than vinegar if I find no inventors that acquire this book are interested in following through with my services after reading the book

The Customer's Best Interests

I should point out that my marketing philosophy is not "Give the customer what they want." My marketing philosophy is far more restrictive and that is "Give the customer what they want, but only if it is in their best interests." True, we may have some disagreement about what is in your best interests. I would want to thoroughly discuss that with you before making a final decision. It may very well be in your best interests to waste $2,000 or $5,000 on something I expect will be a failure if you have the resources and don't want to be forever plagued with the "What wouldda happened if?" question. On the other hand, you can readily find people who will give you that positive, warm fuzzy feeling that you crave for your idea. **Hold your**

judgement on the sincerity of your paid advisors until your bank account is finally empty though.

Please feel free to contact me via e-mail BUT be aware of the ground rules included in this section. I debated for some time about providing for direct e-mail contact and decided to go ahead and do it on the supposition that I could always cut it off later if it proves to be a bad idea. It will be considered a bad idea if people do not follow the rules or expect me to bend the rules just for them.

Please don't think I am alone in wondering if providing e-mail access is a good move or not. I see far too many questions in my e-mail and on news groups, etc. where IT IS ABSOLUTELY CLEAR THAT THE QUERIER DID NOT DO ANY HOMEWORK. They obviously believe their time is more precious than that of the people who will, just for free, give them 300 pages of advice specifically tailored to their (carefully undisclosed) invention. "How do I get a patent?" "Do you buy ideas?" "I'm about to finish the rules to a new stadium game, who do I sell it to?" That is a partial list from a recent week. You don't want to be categorized as one of them and I don't want to have to categorize you as one of them—I know you are smarter than that.

Compensation IS Expected

The first rule is I do not work for free. Any IDEA you send me without prepaid compensation or a written alternative will be considered a gift for me to use in any way I see fit WITH NO COMPENSATION TO YOU. If you already have secured for yourself some form of intellectual property protection then I can't violate that. If you have no secured protection then consider your IDEA published, your 1 year U.S. patent application clock ticking, and all foreign patenting rights possibly foregone. This may also be what a court would decide happened when you broached your idea to anyone else without a non-disclosure agreement or at least letting them know that it was confidential.

If you submit an IDEA for which you are not the originator, you are solely liable to the originator of the idea for any compensation the originator feels is due regardless of whether the submission was intentional or accidental or whether you believed you were authorized to release the idea or not. If that is not clear, please do not submit any ideas because I fear we will have an ongoing communications problem. If that sounds harsh remember that inventing is a BUSINESS game. It cannot (usually) be successfully played by hobby rules.

You might think a simple policy of return shipping products or refusing letters which are unsolicited or unpaid for would be the simplest and most customer friendly, and the one that will keep me out of legal hot water. The problem is if I don't know to write "Refused" on a package or letter before opening it I'm sunk regardless plus I have to pay the return postage.

The music industry has been fighting this problem for years. If you want to submit a song to them you ordinarily must get telephone permission in advance and mark your package exactly as they tell you; otherwise, it will be returned unopened. They check both the marking and the authorized sender list before opening a package. But they are still not protected; if they refuse a package the sender can still tamper with the package and sue, claiming that the music publisher took out his hit song IDEA, stole it, and returned the "opened" package.

At some point you have to trust people. But I have lost patience with people who; 1) want all the answers free first, before they decide if they might pay, 2) can't understand that other people might just be smarter or have more knowledge or expertise in an area than them, and 3) fail to see that I can make a far better living helping them get rich than by stealing even unprotected ideas and battling their lawsuits.

If you have an idea or product you want evaluated, send, 1) the check, 2) a solid, comprehendible description (and possibly drawings) of your invention idea, and 3) your non-disclosure agreement (if you wish) to the address in the front of this book. Any non-disclosure agreement that prohibits disclosure to prospective customers or others that might be needed for proper evaluation will be modified, replaced, or rejected. If you are really paranoid, you can send the check and a generic non-disclosure agreement, but be aware that for any generic non-disclosure agreement to have significant value you might need to subsequently provide specific non-disclosure agreements with anything you send.

I reserve the right to modify or reject outright any non-disclosure agreement that I find unacceptable. I will return your check and the rejected non-disclosure agreement (and attachments) or hold your check until you ratify any revised non-disclosure agreement. The intent here is to work out a non-disclosure that is acceptable to both of us. The bottom line though is

James E. White

that other people have to know what an invention is before I, or anyone else, can collect information on whether they would buy it.

The Offers

All prices are, of course, subject to change without notice or reprinting or revisions, etc., of this book.

The least expensive offer is that I provide a quick and dirty report for $100 and requiring no more than 2 hours of my time. The report will include:

- a dirty, online patent check (unless you specify not to) which is not to be confused with a bona fide Patent Search,

- a quick and dirty Step 0, See What's Available, check for other products that solve the same problem, and

- a gross market size estimate from generally available published market statistics. (Specialty products may not have generally available data and therefore will get a best guess and how it was arrived at.)

Remember you get 2 hours of results, expect no more. No opinion on salability will be provided. The "report" will be a letter.

A more moderate level offer is that for $2,000 plus Patent Search costs (typically from $300 to $800 and done by a registered patent attorney or agent, not me) I will provide a much more detailed report including:

- the Patent Search result,

- a WIN submission and report,

- a Step 0, See What's Available, check,

- a suggested retail price range based on competitive products (if any),

- a gross market size estimate,

- a listing of suggested advertisable buyer benefits and matching product points,

- a suggested first draft package design and copy,

- a suggested first draft advertising layout and copy with at least 5 possible headlines,

- a first draft press release for product introduction or patent award (whichever I guess will occur first for your product),

- a list of publications and their addresses (with unverified contact names when available) to send the press releases to,
- my opinion on salability, and
- my opinion on your business risk.

If all that takes 10 hours or more to pull together, your effective hourly cost will be $200 or less which is very reasonable for professional support. After less than $1,000 of my effort, if the Patent Search or WIN results do not look promising I will contact you to see if you prefer a refund of the balance remaining or to continue with the research and report. The above essentially includes the marketing aspects of Steps 0, See What's Available, and 2, Marketing Evaluation. The report will be a cover letter incorporating what was learned plus each of the items noted above.

For an additional $2,000 I will do:

- the marketing parts of Step 1, Asking People, including experts, about the invention.

The effort will either be via personal interviews or a focus group or groups. You will be responsible for providing a model or prototype (or at the very least, drawings) should they be needed in explaining the invention or helping respondents visualize the invention. A brief report will be prepared and will have all responses attached in an appendix. Be aware that you will be responsible for any sketches or engineering drawings. I have contacts that can do them for a fee but I do not do them myself.

For an additional $2,000 I will do:

- the off-the-shelf products divided by ten production cost analysis, plus
- put together a quote request package and contact at least 3 manufacturers to try to get quotes.

If 6 manufacturers are contacted and they all quote "no bid" then that is the result you get. (Such a situation should be extremely rare for products using standard materials and techniques.) The report will be brief and include the results and what the comparable products were and who the manufacturers were and their bids.

For $10,000 <u>per month</u> I will provide about 40 hours a month writing copy for ads and directing you and your staff (or third parties at your expense) in what to do to pull your marketing together from, getting your

UPC to placing your ads (at your expense). This may include identifying the best current marketing channel or providing insights on alternative marketing channels that may be better.

While $10,000 may sound high, that is a professional services rate of only $250 per hour, but you should be able to dramatically leverage that by having your own less expensive support staff do the brunt of the work. Also, the odds are that you will only want (or need) to buy one or two months of such service to get the marketing effort underway to sell millions of dollars of product. That is in direct contrast with marketing firms that for higher fees (but possibly lower rates) will want to take all the marketing effort in-house and complete it themselves.

I will give you many decision points along the way where you must decide if I or you do the work or if I or you locate a contractor to perform specific functions such as letterhead design or printing or media buying.

The most complete service will be provided for a negotiated deal where I handle all the marketing work (and optionally production) completely at my expense and I get a (substantial) part of the profits. This, of course, takes the risk completely off you and puts it on me. I do recommend that you exercise at least the $2,000 offer first; otherwise I reserve the right to simply reject your invention with no further negotiation and without providing any research or report or reason. In other words, I hope I am not stupid enough to negotiate a percent of "profits" for an invention which will (probably) never be profitably marketable. So what if I might occasionally overlook a winner that you should pursue—you didn't believe in it $2,000 worth either!

If you read through the list of offers above and thought of them all as "final," you are not thinking BUSINESS yet. In the business game things are negotiable. You can ask for exactly what you want and then negotiate a price that is acceptable to both parties. Just a reminder: my quoted negotiating starting prices above are subject to change without notice.

Yes, I know, starting negotiations is tough. If you just pony up the stated price (as possibly revised) I'll accept it with open arms and give you what you pay for. If you ask for some changes in effort or price THERE IS A CHANCE I'll reject your proposal (not YOU)—and that might not feel good. The chances in my case—and most business cases—of proposal rejection are a lot smaller than you think. In business, the person that is thought of as "funny" is not the person who asks for something and is rejected but the one who is rejected AND DOESN'T COME BACK WITH A NEW PROPOSAL when it is warranted.

Marketing Often Looks Easy

Be forewarned! Marketing your product will probably look "easy" after you've seen it done. Since I work with clients on a percent of increased profits basis I always tell them to keep their current marketing going while I get started. They nearly always drop their own advertising in 2 or 3 months since mine is obviously pulling in more clients or customers.

Because of that, my agreements also call for them to make payments equivalent to their current advertising payments directly to me should they drop their advertising. Soon enough they also realize that they can write ads with the techniques I have introduced and they decide that rather than pay me for my ads they'll just write their own. That is why my agreements say any use of the techniques, offers, etc., that I introduce, whether I wrote the ad or not, still obligate them to pay me a share of the profits.

Proper marketing (for most products) should get measurable results almost instantly. If you have put a product on the market and are not getting results within 1 month, 99% of the time either you are doing something grossly wrong in your marketing or your product isn't viable in the first place. Maybe 95% of the time it will be the latter since **"A product that won't sell without advertising, will not sell with advertising"** to quote Albert Lasker, a famous advertising/marketing man from not too many years ago.

Guest Speaking

I would also be happy to provide a talk (1-2 hours) at your inventors group meeting provided that all my expenses are paid to do so (travel, food, lodging). You will get the time for free and I take the chance that at least one of your club members is serious about getting a profitably marketable invention to market with a little (of my) help. Of course, I also want the right to sell books at the meeting. Business is business.

James E. White

CHAPTER 13

Doing Your Own Patent Search

I highly recommend you do a search yourself and that you start it on the www.uspto.gov site. Caution: in the more than one year (Wow, has it been that long?) that I've been working on this book there have been numerous changes in the way online searching is done. In other words, read this material for its general content and process and don't necessarily expect each detail to still be correct. In fact, the most recent changes before printing of this book occurred November 19, 1999, just 3 weeks before it went to the printer. Classification Index searching was (kind of) added.

Find All Probable Classification(s) First

The current URL for classifications is www.uspto.gov/web/offices/pac/clasdefs/index.html but you can get to that from the USPTO home page by clicking "Site Index" then "Classification Definitions, Patent" or by clicking "Patents" then "Patent Class Definitions." Read through the entire list and write down the ones you believe might be relevant.

After reading down the list for a few seconds your first thought will probably be "There ought to be a way to make sense of this." Sorry, there isn't. If there was do you think all those minions that keep taking your patent processing money would have jobs? Even at the deeper levels there is no inherent sense to it. **Treat it like an art.** If you want some clues you can look at *Examiner Handbook to the US Patent Classification System*, particularly its Appendix A, which is available from the "Patent" button on the USPTO home page.

Next click on one of the relevant choices. For illustrative purposes I'll choose "4: Baths, Closets, Sinks, and Spittoons" to look for toilet seat lifters. I'm cheating of course because I already know that "closets" in patent speak, in part, means "water-closets" which means "toilets" in modern U.S. English. You may be lucky and hit on "63: Jewelry" and have exactly what you want with no interpretation needed. My choice dumps me into a huge page for "CLASS 4, BATHS, CLOSETS, SINKS, AND SPITTOONS" and here we used to see just how easily "good" ideas go bad.

The published "printed" document that you are now viewing online uses indents to distinguish groups and sub-groups. The online document does not have the indents and therefore, until November 19, 1999, was very difficult to impossible to use. Now the document has wonderful line gaps between sections and subsections and has color headings and subheadings for classifications. In short, it's a major improvement.

Original HTML as a Lesson

The printed versions of USPTO, and many other publications use indents and tabs for easier reading. Unfortunately the original HTML (Hyper-text Markup Language) creators either hated leading spaces and tabs or they didn't have the brains to deal with them correctly or they (mistakenly) thought getting rid of them was a great idea. Your choice, (I'm informed the latter is correct) but be aware that your invention will be critiqued by the same 20/20 hindsight that millions of users now look at HTML results with, regardless of how "right" your reasoning is.

HTML rules currently severely frown on use of "tables" to achieve a formatted appearance yet virtually 100% of all web sites resort to tables to achieve eye pleasing formatting. And only recently have non-backward compatible standards for "style" control (e.g., paragraph indents) been added to HTML. (Remember, "What do people want?")

First read the class definition and its notes. There may be too many notes to easily comprehend. The preliminaries in "Jewelry," for example, are less than a page, while the preliminaries in "Metal Deforming" are probably over 20 pages. In any category, all the preliminaries end before the word "SUBCLASSES" capitalized in red at the left margin. If nothing clearly jumps out at you from the preliminaries as to where you need to go to locate the correct class/subclass for you, you'll need to do a text string search.

Hold the "Ctrl" key on your keyboard down while tapping the "Home" key once to move back to the beginning of the page. Choose the "Edit" choice from the menu bar across the top of your browser, then chose "Find." Enter a text string that you think will find the right subclass for your invention. The string should not be too specific or you may not find anything and it should not be too general or you may have to wade through too much. Also make sure that you do not check any boxes that limit your search to exact capitalization (case) or whole word matching.

James E. White

I tried "seat lift," "lifter," "lifting," "seat raise," and several others without finding a single instance of my search string until I tried just "lift." [Note: my editor insists we follow the "rules" here and include the punctuation within the quotes even though I'd prefer to put it outside the quotes so it wouldn't confuse new computer users. In other words, search "seat lift" rather than "seat lift,", okay?] Unfortunately everything for "lift" was not relevant so I finally resigned myself to searching for "seat". It took 25 "Find Next" clicks to come across "246.1+, for an opener or closer for a closet seat or lid." as a "SEARCH THIS CLASS, SUBCLASS" instruction under subclass 239. From this I see that "opener" is the patent office terminology for what I want.

I closed the "Find" box, did "Ctrl-Home" to the top, and "Edit/Find" for "opener" and in 2 "Find Next" clicks I end up right on "246.1 OPENER..." Reading down through the 246 subclass I come across "246.4 Four bar link:" with a paragraph of text description that sounds exactly like what I imagine my inventors club friend's "invention" to be. Wow, the wording may be arcane but my friend's exact "invention" already has a sub-subclass and he just invented it! For my own invention I was easily able to locate 63/35 "Jewelry Findings."

While you are still online you also should visit the Manual of Classification at www.uspto.gov with its "dot" hierarchy. First click "Patents" then "Manual of Classification" then, when the list of numbers with titles finally appears, look down the list to find the proper class or potential classes. I chose "004 BATHS, CLOSETS, SINKS, AND SPITTOONS."

Now look down the displayed terms reading ONLY THE ALL CAPITALIZED ones because they are the highest level of the dot hierarchy and any terms at a subordinate dot level are only applicable if the invention properly belongs WITHIN the higher level. For a particularly long list you might try **Edit|Find** from the browser menu, being sure to type in a significant word in all CAPITAL letters and checking the "Match case" checkbox. Only if an upper case search doesn't find it should you switch to lower case searching. When a lower case search finds something, read back up the dot hierarchy and see if the first upper case category above your find is appropriate; if it's not, keep looking.

Unfortunately, the USPTO does not give you indexed keyword access to the classification hierarchy online. A site that predates the USPTO's online classification access, and does give index access, is "Index to Manual of Classification of Patents" at metalab.unc.edu/patents. Unfortunately this site is out of date and may disappear at any time. Try it and if it is still there you can use the following search as a guide to using it. [As of November 19, 1999 the USPTO does provide some alphabetic access to the classification index but you have to guess the right word to get into the middle of the hierarchy. By the time you read this I hope they have full keyword searching available.]

To do my search at "metalab" I chose "TITANATE to TRANSURANI-UM" from the alphabetic list that lets you first narrow your search. The word "toilet" is alphabetically between those two terms. In looking down the retrieved index page "SEAT ....4/234+" looked like the best choice. Next I modified the URL entry/selection line in my browser to go to metalab.unc. edu/patents/class for the "Index of /patents/class list." (I just "knew" that URL was there.) This list is totally alphabetical so 4 falls well after 100, for example. I scrolled all the way down to "CLASS4.html" and chose it rather than "CLASS4.txt" which would do just as well. That got me to the dotted hierarchy index itself, just like you would see it in the print or USPTO online version of the *Manual of Classification*, except that online here you're limited to a small section of it.

Next I wanted to find 234 because it is the subclass number located previously. At the top of the list the numbers start with 600s then below that 300s. What order is this list in anyway? (None, at least none discernable, is the correct answer.) Scrolling on down shows 300s, 400s, 500s, 600s (again but different), 100s and finally 200s. There it is, "234 SEAT AND LID" but the dot entries under it don't quite look like what I want.

I keep reading down and still seem to be in toilet seat territory. When doing this, particularly pay attention to the ALL CAPITALIZED words because they are the uppermost level of the dot hierarchy. In this section you'll see "SEAT" and "LID" and "COVER FOR SEAT" before coming to:

246.1 OPENER OR CLOSER FOR A CLOSET SEAT OR LID

246.2 .Fluid mechanism

246.3 .Lever operated opener

246.4 ..Four bar link

246.5 ...Using a seat or lid for a link

Bingo! 4/246+ is obviously the class/subclass area we want for our toilet seat lifter. That was pretty easy... or was it just lucky?

Trying jewelry (at metalab), "Class 63" is easy but no subclass seems to fit my invention. Going to the "/class" page and finding CLASS63 and starting down from the top we find "2 MISCELLANEOUS" right away. If you find such a category you'll need to keep track of that because you'll need it if nothing specific is a direct hit. Even with a couple of readings of the list nothing stands out as correct but there is a "21 WATCH AND CHAIN ATTACHMENTS" which might be relevant if we call a necklace a chain.

"Findings" does not appear anywhere in the list. This kind of search failure in the indexes is why I recommend you do string (**Edit|Find**) searches in the Classification Definitions of the USPTO site first, even before visiting a patent library. (Doing a string search for "finding" in the "063 Jewelry" section of the current online Manual of Classification does find it, but as subclass 35 under 33 MISCELLANEOUS.)

Look at Patents to Be Certain

Next go to the patent database itself and do a boolean search. (From the www.uspto.gov home page choose "Searchable Databases" then "Patent Full-Text Database with Full-Page Images" then "Boolean Search.") You'll probably find searching a bit slower than if you chose the Bibliographic (and Abstract) database. But, if you used the Bibliographic database, you would discover that you wouldn't get to see anything useful enough for you to determine how close your idea is to an existing patent. Titles are often totally meaningless and abstracts either so vague or so arcanely mechanically descriptive it is almost impossible to imagine what they mean.

In the "Term 1" search entry box enter the class number then a slash then the subclass number (e.g., 4/246.3). Select "Issued US Classification" from the "in Field 1" list and select "All years" from the "Select years" list. Hopefully you have some "hits" (computer nerd lingo for "found results").

Systematically go down the list selecting the ones that seem most relevant and look at the class/subclass numbers they have listed for U.S. Class and Field of Search. Write down the different class/subclass numbers and keep track of how often they occur. The ones that are close to your search may not tell you anything (e.g., 4/246.1, 246.2, 246.3, 246.4, 246.5, 248, 250) but the ones that are different (e.g., 4/408) might be useful to look up, perhaps both as a database search and a classification. You may turn up some other area that is better suited to your idea than the one you found at first. Or your invention may truly cross areas, which makes your searching effort harder of course.

The www.patents.ibm.com site can also be used for the above kind of search but be aware their field name is "U.S./National Class." It is unfortunate that neither IBM nor the USPTO allow searching of the "Field of Search" information on a patent front page since this would be the ideal way to cross-check your Class. The IBM site does provide cross-references by U. S. References (i.e., U.S. Patent Prior Art references) which can be useful. You often won't be able to see the referenced patent itself but you will be able to see some of the other patents since 1971 that also reference it.

While you are still online you might just want to poke your nose into the www.uspto.gov/web/offices/ac/ido/oeip/taf/reports.htm page of "Statistical Reports Available for Viewing" which among other things will let you access the "Patent Counts by Class by Year Report" and the "Patent Counts by Class by Year, Independent Inventors" tables. Looking at these is not mandatory but it will give you some idea of the activity level in the class for your idea. A very active class probably means there is money to be made IF YOU'RE THE WINNER while a relatively calm class might indicate clear sailing or no consumer interest in problem solutions.

Visit a PTDL

Your next step is to visit your nearest PTDL (Patent and Trademark Depository Library). To find it select "PTDL Library List" from the USPTO Site Index. If you are lucky you will also find a link to that library's web site where you can find the hours they are open (often evening hours at least one day a week), directions, and parking information. At worst you'll get only a state and phone number (keep in mind the nearest library for you may be in a state bordering yours). Also an extra hour or two drive to a Partnership Site or one with online APS Text Searching capability may be well worth it. Call and ask about fees before you go since some sites charge to recover telephone and printing costs.

When you get to the library find the PTDL section, ASK THE LIBRARIAN there for help. Your tax dollars have already paid them to help you so you need not be bashful. They expect you to be IGNORANT (of the search process) so go ahead and act like it and you will learn a lot more and have a much more productive experience. They also expect you to be INTELLIGENT enough to ask questions when you don't understand or miss something they say.

James E. White

Plan, by the way, to spend a minimum of 4 hours but more typically 8-12 hours doing your search over a couple of days. While you are working there you will undoubtedly see "KNOWLEDGEABLE" (note the quotes) and STUPID people (obvious wannabes) that bounce off the proffered help, spend 15 minutes to an hour shuffling about without seeming to get a handle on anything and leave. Their trip to the library was a waste of time. The librarians will NOT do your search, they will provide generic guidance, and they will NOT answer legal questions or provide any opinions on patentability or your invention versus another one, etc.

The Index

The first thing they will guide you to is the *Index to the U.S. Patent Classification System*. Some sites may have this index (and other books) available for sale or you can find it at the U.S. Government Printing Office (GPO) site at www.gpo.gov. The easiest way to find it there is to choose "GPO Access" "Finding Aids" from the GPO home page then chose the "Catalog of U.S. Government Publications (MOCAT)" database and do a keyword search for "patent classification index." When I last looked at the "Short Record" the annual price was $22 ($100 on CD-ROM).

If you get really interested in spending money to buy your own copies of Patent Office documents you can choose "Classification Publications" or "Catalog, Products and Services (CD-ROM, tapes, etc.)" from the USPTO Site Index. Rummage around a bit and you will find that for $1,200 you can buy CD-ROMS of U.S., European, and Japanese patent first pages since 1971 and you can subscribe to the updates.

Keep in mind the index is only a starting point and not a definitive answer. You may discover, as I did, that the exact term you want, and that you found in your online classification definitions search, is mysteriously missing from the index. In my case "findings" or "jewelry, findings" was not there. Is this an honest mistake, a misguided attempt to use only "natural language," bureaucratic incompetence at its "finest," or something else? Even though the Patent Office is NOT tax supported, be aware, in the patenting process you are entering the "government zone" where reality is not quite the same as... well, reality.

If your invention is for a specific part of some larger item try looking up both the part and the whole. If your invention or the whole might go by several different names, look up each. Write down the Class and Subclass numbers IN PAIRS from each of the things you find that might be close. You will be better off with too many than too few. Remember, while the words in the index look like U.S. English words, you don't yet know what they mean in patentese.

You may come across classes beginning with "D" for Design even though you know you want a utility patent. Go ahead and write them down and plan on looking at them. Also think about your invention or improvement and really determine whether it is truly "functional" or merely "decorative." A new edge for a cookie cutter may be functional while a new shape for a cookie cutter is decorative and can only receive a design patent.

Classification "Dot" Hierarchy

After mining the index for all you think you'll find ask for help on the next step. It will probably be a search by class and subclass in the *Manual of Classification* which is in 2 parts (and isn't a manual at all), one for design patents and one for utility and plant patents. Use the correct book and find the beginning of the class you want to look in.

WARNING: Your instructor will probably tell you but you won't notice, the SUBCLASS numbers are NOT necessarily in NUMERICAL order. **If you try to find your subclass using numeric sequencing as a rationale from within the middle of a 20-page class, you will likely get frustrated.** The librarian I asked about it had no idea why a sound numerical sequence was not followed with cross-references where appropriate.

Be extremely careful with the hierarchy in this document and, if a higher level of the hierarchy is clearly not applicable for your invention relative to the subclass number you found earlier, cross out your entry for that class/subclass pair only; the class paired with another subclass may still be applicable.

Also be aware that "Miscellaneous" or other categories to catch whatever doesn't have a specific subclass or lower hierarchical level may be at the very end of the hierarchical "dot" level you are at and may not be on the current page. **It is not unusual to find new, more applicable, subclass numbers than you found through the index.** Don't forget to write them down. The hierarchy should do a fairly good job narrowing down your relevant class/subclass pairs.

Check the Definitions

Ask for help once again and you'll either get directed to an online or a paper (or equivalent microfiche) version of the Classification Definitions you looked at online via the Internet. WARNING: The online (CD ROM) version the library will have you look at is NOT browsable (at least from what I've seen). I highly recommend you ask to see the paper version if at all possible and the microfiche version if that fails.

The reason you need the full text is the index and dot hierarchies you looked at were NOT exhaustive, meaning the full depth of the classification hierarchy is NOT included and, worse, whole groups of subclass numbers are OMITTED ENTIRELY. Particularly prone to this kind of omission are significant, unrelated, "Miscellaneous" sub and subsubclasses. Since you only look at definitions online by searching the class/subclass numbers you have written down you'll only be able to see if what you've found so far appears to be the right area. [Also see the Digests and X-Art Collections a few pages farther on in the book.]

You have almost no chance of finding the exactly correct area if it was omitted from the index unless, somewhere in the FULL definition, you happen on a note that directs you there. That kind of omission is why "finding" (for my jewelry search) did not appear in the index. Look at the FULL definitions and any notes associated with them. If you must look at the definitions online, make sure the librarian shows you how to see the FULL definition and ALWAYS look at it even when it appears that the short version is complete. The short version of the definition may just have paused at a blank line that precedes the gold nugget note relevant to your invention. Now it gets tricky.

Some definitions should be taken literally while others should be interpreted. Why? Because at any instant in time the classification definitions are considered complete and static. Over a long period of time, of course, they can't be. Truly new inventions like electronics and televisions and computers crop up. They can be fitted into "the" classification definitions for a while but it soon becomes obvious that something in the classification scheme has to change.

So a new class or subclass or subclass group is born and its definition is "precisely" written. The new definitions are easy enough to interpret for anyone knowledgeable in the field of the class/subclass(es) at the time. Time passes. "Hot" subclasses get used constantly, "cold" classes languish and the (patent office) people knowledgeable in them drift away. A fairly new invention arrives, it gets assigned to a fairly new examiner, the class it belongs in is obvious but the subclass is uncertain.

The new examiner may have a college engineering degree but has never worked in the jewelry field. The new examiner also knows, or soon will be taught, that they are a peon, just like their seniors before them have been. As peons they cannot even suggest that a new class or subclass might be needed. It is but an instant in time and they must fit the new into the current and "precise" classification scheme. Slowly they read down the subclass definitions in the appropriate class.

Just as they catch themselves nodding off they see it, "21. WATCH OR CHAIN ATTACHMENT." It doesn't matter that "watch" is the predominant word in subclass 21 and its related subsubclasses 22-25. The literal word "CHAIN" and the words "attachments for" regardless of other language saying "through a buttonhole" or "watch" are literally exact for a necklace (chain) attachment.

The examiner also knows that necklaces, though clearly belonging to subclass 3, did not get specific subclass groups like bracelets and rings did (it's not his, or your, job to wonder why). But the "precise" definition written in the memoryless past did not say "Watches and watch chains" or "ornamental and safety attachments for watches and for ornamental watch chains," the underlined word "watch" was omitted and it just said "chains." The new necklace (chain) attachment inventions get literally fitted in amongst the watch chain attachments.

If the new examiner had chanced to read further they might have eventually come to "33. MISCELLANEOUS" and one of its subsubclasses "35. Findings" but the definition there would give no indication of exactly what a finding is or what items a jeweler would call findings. Our new examiner/engineer would also be unlikely to know that jewelers, and cobblers, and dressmakers, all have "findings"—and that they are all generally different items. This particular meaning of the word "findings" doesn't generally appear in pocket or desktop (abridged) dictionaries.

If you have a sharp memory you may recall that 63/2 was "Miscellaneous" in our (metalab) online *Manual of Classification* search, not the 63/33 we see here. Keep your eyes peeled for these kinds of discrepancies in your

James E. White

search and don't hesitate to keep both class/subclass pairs in your list for subsequent patent searching.

The specific subclasses and terminologies above are real but the exact scenario of the new patent examiner is, of course, made up. It is, however, not unlike the reality of the way any classification scheme from insects to psychological troubles to library subject catalogs gets created and maintained. If you are expecting magic and total clarity at this instant in time, it just isn't there and never will be.

Exact and correct classification schemes, such that anybody (or everybody) would agree are correct, just don't exist. To create and maintain them is exhausting and labor intensive in itself. To imagine that they could be totally retroactively applied is almost beyond belief, even with the aid of computers. Any retroactive application to existing printed documents is, of course, impossible. **By now you're probably beginning to see that an expert searcher, providing that they do a conscientious job for you, earns their fee.**

Digests, X-Art Collections, and Change Orders

Sometimes you will find a class/subclass pair in a patent or in the *Gazette* or even in the indexes but when you go to look it up online you get an "invalid classification" or it doesn't appear in the printed definitions. You can double check to be sure you didn't write it down wrong but what may be happening is you are discovering that the U.S. patent classification "system" is dynamic. It changes all the time and not all documents are (or will ever be) current. In 1997, 417,886 patents were reclassified due to classification system changes. That's about 3 times more than were issued. Ask your PTDL librarian for help and you will most likely be directed to:

1) an (unofficial) "Digest" created by an examiner to group related patents/subclasses that the existing classification structure does not group, or

2) an (official) Cross-reference Art Collection (X-Art) based on a concept other than "proximate function" that groups things on a logical basis even when the grouping doesn't strictly fit the classification rules, or

3) *Classification Change Orders* which are notices of changes to the classification system and/or the definitions..

One of these will likely get you back on track or turn up other areas that might warrant your attention. For example, if you invent a new superconductor material you might want to also research processes for making other superconductor materials. While X-Art references may appear at the end of a class definition, the other items won't. When you need the librarian's help, ASK FOR IT, you already paid for it so get your money's worth. Ten minutes with even an unofficial examiner Digest may save you thousands of dollars.

Retrieve Patent Numbers

Now you can take the class/subclass pairs you have left (hopefully it's less than 3) and ask for help with the next step. You'll do multiple computer searches and end up with patent numbers and titles for utility patents since 1969 and just patent numbers before that all the way back to 1790. If your invention is in the computer field (or some other recent field) you may not need to do pre-1969 searching but it won't hurt except for the time it takes.

From a searching and getting a patent perspective, the best scenario is that you only come up with a few to a few dozen possible patent numbers. From a marketing perspective, an area with a huge number of patents, particularly if most of those are recent and assigned to large firms, means you're in territory in which there is money to be made for those with the best products and competent marketing and legal support.

Check the Gazette Entry

With the list of patent numbers (and titles) in hand you are ready to ask how to use the *Official Gazette*. It may be a printed and bound version or a microfiche. If the title on your list seems to clearly indicate that the invention is not prior art for your invention, you can skip looking up the patent number in the *Gazette*. Conversely, if you clearly see from the title that the patent is relevant to your invention, you can also skip looking it up in the *Gazette*. Huh? The choice is yours because it depends on what you are looking for.

Are you looking to see if your invention specifically and completely is already invented or are you looking to examine all relevant prior art? If you just want to see if your invention is previously invented then the abstract and sketch in the *Gazette* are likely to be sufficient; otherwise, you'll have to look at the patent anyway so there is no point in wasting time looking in the

Gazette. If the title is inconclusive and you want to examine all prior art and searching for the full patent would be done (slowly) on microfiche or (expensively) by paying for computer time, then checking it in the *Gazette* might be worth the time to rule a particular patent number in or out.

Look at Patents

The final step is looking at the patent either on microfiche or on the computer. Ask your librarian for the options and the costs, including printouts. If you have hundreds or maybe even just dozens of patents to look at the chances of their being sufficient "newness" in your idea to warrant the expense of patenting and sufficient "newness" to win customers over from the competition may get pretty slim. You will also quickly decide that those $250 deals for any patent search really sound pretty good. What you may not realize is the cheap deal usually does NOT include copies of the patents and their "standard" price per printout is $4.00 (about twice what their wholesale contract price is) and that the searcher DOES NOT examine the patents or advise you on what conflicts with yours and what doesn't. Read the deal carefully and don't hesitate to ask questions about it.

Attorney and/or Engineer Review

A lawyer or lawyer and engineer, of course, in addition to giving you the found patents, can give you some of the careful examination and advice you need for an additional fee. A patent attorney or agent can give you an opinion on whether there are conflicts with what might be the claims of your invention.

You get no guarantee that your application won't be rejected for something in the U.S. Patents they didn't find. The best guarantee I've seen is that you'll get your patent search fee back if your patent application is finally rejected due to the searcher missing something in the U.S. patent files.

It's not much of a guarantee when you realize they can simply give you a big stack of patents to avoid, that by the time you find out something they didn't turn up causes your rejection you are already thousands of dollars past the small search fee, that you can be rejected for many reasons other than just a prior U.S. patent, and that, if they talked you into using their patent attorneys for the application, etc., they probably know how to write a narrow enough patent that it won't be rejected (but it won't be worth much to you either).

Please don't cry. It's not that hopeless. Doing a patent search yourself really is for your own benefit. If you come out of it fairly certain you have clear patent sailing ahead of you that's great. If you come out of it knowing you are flying into the teeth of huge corporations and you'll have to thread your way through a fairly narrow minefield of prior art, then you may just want to think about your next idea and leave this one for later.

What Will the Examiner Find?

The real catch is, just because you've looked at the right category doesn't mean you'll find it all. My attorney cited an example of a patent for an airplane part being denied because it was similar to a patent for a money clip—which the examiner said you could throw so it "flew" like an airplane. The attorney also said the next level up in the patent office would have likely overturned the examiner but the inventor was tired of fighting and didn't want to spend the money and gave up.

A good look at existing patents should be well worth the time; however, don't expect your attorney to be happy with accepting your results. They will probably insist you pay for them or another professional to do the search. My understanding is that examiners love to rub patent attorney noses in the requirement for citing "prior art," especially if something important but not directly categorized is found by the examiner. Patent examiners, after all, are human, they know that their <u>power</u> is only displayed when they say "NO." While it "technically" is not mandatory that you or your attorney ferret out prior art that you don't already know about, I think it is foolish for you not to. If $500 of searching finds your invention in the prior art that saves you from spending $3,000 paying your attorney to write a worthless patent application. If you spend $6,700 on a patent that goes all the way through to being granted your $6,700 is wasted the instant someone else turns up your invention in the prior art because your patent is (and always was) invalid.

Besides, even if you do catch everything yourself 3 out of 4 times, the extra cost of the patent attorney doing it could easily be made up that 4th time. Again, be aware that patent searching is an art, not a science. For the jewelry invention I am currently working on, my own search, the Wisconsin

Innovation Service Center search, and my patent attorney's searches all turned up different, but relevant, "prior art."

If you do find something in your patent search that looks like it might be a competitor to you, try to find it on the market: you may have missed it in STEP 0. If it isn't there that may give you a clue about weak (or no) demand or, very likely, the inventor didn't have the money or the drive to commercialize it and no one jumped at the chance to license it.

Does your invention avoid infringing on the competitive patent(s)? Can you engineer around infringing? Can others solving the same problem engineer around you, too? If you are uncertain of your patent search (assuming you did one) in STEP 2, that may be the first point in working on your invention that you should spend real money to have a good professional patent search done. It will pay to carefully answer the above questions and evaluate your risks.

If you can be easily engineered around such that a patent won't give you any significant lead time on the market you may want to consider not bothering with the trouble and expense of patenting at all. The major risk then is that someone else invents and patents your invention and you cannot get it invalidated because you don't have documentation to prove your invention was "published" or product on the market 1 year before the "someone else" filed their application or you can't find the evidence to show that they falsely claimed "inventor" status after seeing or otherwise finding out about your invention.

The time window, assuming "diligent pursuit" on your part, for such an unlikely scenario is very small and, at worst, your cost will likely be just engineering around them unless they have the finances to harass you with extended litigation. The original "childproof" cap was invented by a local club member; Ralph Nader then got Congress to mandate childproof caps. While the inventor got the patent he only sold to one company, all other manufacturers chose to copy "prior art" (radiator caps) or engineer around the patented invention's claims. Everybody benefits from childproof caps and the inventor is certainly pleased with that.

If you are developing in a "hot" area (such as electric lamps in the 1870s) where the basic patents may not yet be established or filed, you will, of course, always want to go for a patent. It may cost you an arm and a leg just to get beat out by an interfering patent. Deal with it as a rational BUSINESS gamble.

CHAPTER 14

Resources

This chapter first adds a few resources then duplicates a few but adds some comments about them. The chapter ends with an index that lists all resources mentioned in this book and shows where they are referred to in the text. The resources here are ones I've found to be particularly useful but, by that, I do not mean to imply that resources not listed are therefore not particularly useful—it is probably more likely that I just have not seen them. The items are listed in title order so no priority is to be understood from their sequence.

Books

A Streak of Luck, Robert Conot, 1979. A very good chronicle of Thomas Edison and his fantastic inventive streak. The book provides an excellent look at the times Edison lived in and the swirl of ideas being frantically worked on by many. Seven years, and more than $100,000 in litigation expenses, after Edison's patent was invalidated by the Patent Office, on October 6, 1889, a judge ruled that the electric light improvement claim for "a filament of carbon of high resistance" was valid.

The book, unfortunately, is also able to clearly show that Edison and his attorneys hid significant information from the judge that made that determination. They cut out the October 7-21 section of a notebook that the judge might have determined showed that they were simply extending Sawyer's (or Swan's) work with carbon "burners" or "rods" in an evacuated glass bulb. Now you know why bound inventor's notebooks MUST have consecutive page numbers! The reality probably is that all Edison and his team did was change the name to "filament." They, in fact, did not find a commercially viable filament (bamboo) until more than 6 months after filing the patent application. In 1906 the tungsten filament laid the carbon filament to rest for good.

Bringing Your Product To Market, Don Debelak, 1997. A pretty good book that covers all the bases but, in my opinion, just doesn't have the oomph

James E. White

to kick you in the teeth when you need it. The author specializes in assisting inventors in bringing their products to market.

Creating Demand, Powerful Tips and Tactics for Marketing Your Product or Service, Richard Ott, 1992. Make no mistake, this book is not about manipulating consumers to buy your product, it recognizes right in the preface that the consumer is king and will make their own choices on where to spend their money. That said, this book provides a lot of sound fundamentals that will help you market a product that consumers do find desirable.

From Patent to Profit, Bob DeMatteis, 1997 (www.frompatenttoprofit. com). This is an excellent book that favors licensing based on the author's experience. The kinds of products Bob develops (typically plastic bags and bagging systems) are ones where I would also recommend licensing rather than venturing. The main reason is simple. In the bagging industry you must be able to go from 0 to 1,000,000 units a month on day one or you can't play with the customers that will make or break you. The book has excellent license agreement examples and Bob also tells a nice story about why he now always pays for professional patent searching even though he is pretty good at it himself. The book also includes fuller discussions of "trade dress" and the "Doctrine of Equivalents."

How to License Your Million Dollar Idea, Harvey Reese, 1993. This is a decent little book and even includes a sample license agreement. The author's main experience is in the toy industry where licensing is much more common than in other industries but the principles apply regardless. **Mr. Reese also makes the point that, no matter how much marketing research you do that indicates your product will be a failure, you'll probably ignore the research results—at least for your first invention.** A lot of appendix space is tied up with lists that are kept current online these days and there unfortunately is no list or contact information for toy manufacturers (or agents/brokers) that accept outside ideas.

Marketing Your Invention, Thomas E. Mosley, Jr., 1997. The section on licensing is the best I've seen. There is also a good section on the characteristics of successful new products. This includes the facts that, 1) most products that get joint venture partners or are licensed "have sales," and 2) the inventor had enough money for test marketing (i.e., the inventor took the invention through Step 5-Sell a Few).

Millions From The Mind, Alan R. Tripp, 1992. Lots of terrific examples and practical advice. Keep in mind the author's emphasis is on the winners but he does provide lots of good advice on the steps to take to get there. In my opinion too much emphasis is placed on patenting and not enough on consumer want. Many examples required millions prior to launch also, not that you couldn't do that.

Stand Alone, Inventor! And Make Money With Your New Product Ideas! Robert G. Merrick, 1997. A very good book made even better by a resource section that includes over 800 items. Bob's web site is www.bobmerrick.com.

Magazines

Inventors' Digest. You can also see their web page at www. inventorsdigest.com. This is not the only inventors' magazine; you can find others by starting with the inventor resource links identified below. It is a decent magazine but it is usually mostly fluff articles by consultants who will be happy to work with you for fees. By fluff I mean repetitions of info that you would typically find in your first 2 months of serious research anyway. The most hard-information articles are typically Jack Lander's on making prototypes. Jacks runs the Inventors' Bookstore at www.inventorhelp.com and makes plastic prototypes. Sometimes more importantly than the Inventors' Digest magazine's articles are its industry news and the way it keeps up with legislative initiatives.

Web Sites

About.com, inventors.about.com/education/sciphys/inventors/mbody.htm has lots of links to information useful to inventors and about inventors.

The Capital Connection, The Entrepreneur's Web Site for Financial Resources at www.capital-connection.com is a good place to learn about funding your invention. In particular there is a lot about what YOU WILL HAVE TO DO including business plan development etc. In essence you must earn the respect of a funding source before they will kindly "give" you some money to which many strings are often attached. If your lender demands your house as collateral you'll want to be pretty certain your invention will sell profitably first, don't you think?

One pre-reader of the book checked out the Capital Connection web site, followed a link to www.businessplans.org, and studied a few of the sample business plans there. He found something interesting. What he found was the sample business plan sections that laid out the term of the deal to the capitalist. Several allotted only 5% of the venture to the inventor. He and I both suspect that is much less than you, as an inventor, expect. Since the site is biased toward the capitalists I would look at it only as their starting negotiating point—but don't expect to get much past it if you have no hard cash or serious sweat equity on the table. Which would you rather invest in, an idea or a profitably selling product?

InventNET, The Inventors Network inventnet.com. Of course they start out saying PATENT IT FIRST but their most active contributors are attorneys so there might be some bias there. This site's main claim to fame is a forum where you can ask questions and get answers. The archives of the forum are available without joining but if you want to keep up you'll have to join or subscribe (for free) by following the instructions. Like all Internet forums, it is subject to a fair amount of guess type suggestions and miscellaneous blather, but you can glean solid information if you are intellectually able to separate the wheat from the chaff. (This forum is moderated so it keeps a lot of trash out.) It also helps (on any forum) if you ask a specific question and provide good background.

WARNING: If you ever participate in a forum please read their FAQ (Frequently Asked Questions) pages and rules for participation first to reduce your probability of making (?) a fool of yourself. At this site you can also see a list of what are (possibly fraudulently) called "profitable inventions" which haven't made it out of the starting gate yet (except for patenting costs). You can buy or license the rights to them though.

Inventor's Alliance at www.inventorsalliance.org/rsrcsbm.htm has the Inventors' Resources list which is well worth looking at although it does not seem to be as actively maintained as other sites.

Inventor's Café was a former name for the Patent Café listed below.

Inventors News Group at news:alt.inventors (this is not a URL and NOT to be preceded by "http://") is a news group you can subscribe to (for free) but be aware that there is a lot of chaff in with the wheat. (There is no moderator so you'll see lots of senseless blather and ads.) If your browser is not set up for newsgroups, consult your documentation...and good luck! I suggest that you visit the www.deja.com site and search for "alt.inventors" then you can browse past threads at your leisure or do searches to try to find specific information. You don't have to register till you want to subscribe to the group. You can post answers via Deja also.

The Job Shop Network at www.jobshop.com is a very good place to start looking for manufacturers online. You have to sign up but it's free and then you'll have yet a name and password to remember (Do you get the idea that I think that free sites requiring registration are despicable?). In addition to just looking at the listings of places you should check out their trade show list. You might find one in your area and the trade shows also have links to web sites of manufacturers. The links from the trade shows include many manufacturers that are NOT in their categorized online database.

National Inventors Hall of Fame at www.invent.org/book/index.html The physical facility is in Akron, Ohio. An interesting site but it's geared to the masses who are willing to take dumbed-down information and not be too critical about letting a few facts get in the way of good stories.

Patent Café at www.patentcafe.com also has a lot of information and links and I recommend it highly. This is the inventor/entrepreneur Andy Gibbs' site.

Ronald J. Riley's Inventor Internet Pages at www.inventored.org provides a mishmash of information, editorials, and resources. There is, in my opinion, too strong a bias toward the individual inventor that fails to acknowledge many of the fraudulent "inventors" over the years. You will have to work at it a bit and rummage around to find what you want because the site is very complex and interconnected with occasional dead and misdirected links (even within the site itself). Even if you don't look at his "Why I created this Web Site" page you can't help but be impressed by the amount of labor that went into it. Well worth your trouble.

List of Resources

Appendixes

Sample Agreements, General Information

Each of the following sample agreement appendixes will first have the sample agreement then a section of text with my thoughts on the specific paragraphs I included. Again, I am not a lawyer and none of these agreements have been tested in court or arbitration. Does a lawyer have to draw up an agreement? The answer is NO. In the words of Nolo Press "hold your nose and write." You can have a "legal" agreement just based on an "understanding." The problem with an "understanding" is that the courts generally have no way to determine what that understanding is (or was) believed to be.

The same holds for an oral agreement. If "witnesses" exist to the previous two types of "agreements" then a court might be able to sort things out. In either of the above the courts can only realistically get involved if there is a legal issue to be decided. On the other hand, with a written agreement, the meaning is as clear as the agreement. Obviously this means that whoever does the writing must write as clearly as possible. An "agreement" is just as valid and binding as a "contract," but, I think, has a friendlier ring to it. **Think win-win.**

If you want some help understanding the kinds of clauses you need to include you can try the *Quicken Family Lawyer* by Parsons Technology (start at www.itslegal.com) or other similar software for under $100. More expensive forms software can be had at www.legalstar.com where $1,295 is the full price, but I think the depth is too limited. You can also go the free route and locate www.findforms.com on the Internet then search for things like licensing, non-disclosure, partnership, etc. You can easily be overwhelmed but you will quickly notice that there are some general patterns and conspicuous phrases even though the located materials are far from the same.

Where Lawyer Language Comes From

My attorneys' are always noting that I am not using "standard" contract language and suggesting corrections. Of course each attorney suggests different "standard" corrections. Do they just make up the standards or do they get them from somewhere?

Generally they are quoting words from either "black letter law" (i.e., words specifically included in U.S. and state laws) or from "precedent." "Black letter law" is usually an amalgam of new written law and wording established by "precedent." Precedent (in lawyer terms also called *stare decisis* or *stare decisus*) means that the interpretation of specific words and/or phrases have been decided before by a judicial body to mean something specific that may (or may not be) obvious from the literal text. Also think about how often "precedent" is set by the U.S. Supreme Court where the decision was NOT 9-0 for one interpretation. If 4 out of 9 Supreme Court Justices "don't understand it" what are the odds a jury or an arbitrator will?

And why would judicial bodies have to decide what something means in the first place? Perhaps because it wasn't written clearly enough. Perhaps because some ___hole wanted a specific interpretation for their own benefit. Those are obviously two good reasons to use the "standard" language, i.e., poor wording has gotten a "good" interpretation and ___holes demand what they want. Really interesting.

There is also the little issue of "did the judicial body make the right decision?" A higher judicial body would have to decide that and that only occurs when someone pays money for an appeal. Lots of times on appeal the higher court simply remands the case back to the lower court with the admonishment that the lower court made a faulty decision because they were looking at the wrong law.

"The shear volume of judicial decisions, each possibly creating precedent, makes 'the law' beyond the comprehension of lawyers, let alone laypersons." *The Legal Environment of Business*, Corley & Black, 1981. Pg. 17

From the above quote we see one reason that 2 attorneys use different "standard" language; they use different precedents. Another reason may be differences in state laws and yet another is just their personal opinion of what is clear. I have yet to meet an attorney who couldn't "improve" an agreement written by any other attorney.

Don't get me wrong, I'm a diehard believer in getting the best work I can from an attorney I trust. I'd like to trust that any opposing attorney would have also had the privilege of practicing law "conferred as a result of [their] knowledge of the law *and possession of good moral character* [emphasis added]" (*ibid*, pg. 32), but I'm certain that won't always be the case.

My usual bias is to recommend including the American Arbitration Association clause in your agreements. In general this should get you a far faster, fairer, and less expensive resolution of disputes than taking your case through the "Justice" system. While it is still possible for the big money player in an arbitration to mete out some "punishment" to their "victim," it is likely to be far less than they could do in the courts.

If you do an original of your agreements then make copies, it would probably be best if you designated one person to always keep the originals. One of you or an attorney would be fine. If you do multiple originals each person must be extremely careful to be sure all are exactly the same.

The bold superscript letters (e.g., [A]) in the following agreements match the sections in the notes following the example agreement.

A. Non-Disclosure Agreement - Sample

_____ (Recipient) [A] agrees to
maintain any and all confidential information regarding the _____
_____ idea and/or invention [B] devised in whole or in part
by _____ (Inventor) in confidence and
not to use such information in any way for a period of 5 years [C] without the Inventor's express
written permission. Confidential information does not [D] include: 1) information currently in
the public domain, 2) information subsequently placed in the public domain except by
breach of this agreement by Recipient, 3) information which Recipient can prove was in
their possession at the time of this agreement, and 4) information provided by a third party
provided the third party did not obtain the information directly or indirectly from Inventor.
Any improvements, whether patentable or not, by Recipient shall be owned [E] by Inventor
except as subsequently agreed to or awarded. Patentable improvements [G] by Recipient
must be acknowledged by co-inventor status upon patent application by Inventor. No
license other rights [H] to the idea/invention are granted to Recipient by this agreement.

Recipient and Inventor further agree to submit any controversy or claim arising out of
this agreement, or any breach thereof, which cannot be mutually settled between them,
to binding arbitration [I] to be administered by the American Arbitration Association in
accordance with its rules. Any and all available evidence may be presented to the
arbitrator(s). The arbitration decisions regarding any and all matters will be final.

[J] In the event one party must resort to the courts after the other party fails to live up to
an arbitration settlement, the Recipient and Inventor agree that the judgement on the
award rendered by the American Arbitration Association arbitrator(s) may be entered in
any court having jurisdiction thereof, and that, should the plaintiff party prevail, the other
party will, in addition to the court judgement, pay all costs incurred.

The Inventor will retain an original [K] of this agreement and agrees that the confidential
information is being disclosed to the Recipient for each of the following purposes [L] initialed
by Inventor:

_____ Idea Evaluation*   _____ Marketability Evaluation*   _____ Development   _____ Research
_____ Prototyping _____ Manufacturing or Estimating _____ Development of Marketing Materials*
* Marked activities will require limited disclosures but will be done as confidentially as is reasonable.

M

_____   _____
Recipient Signature   Inventor Signature

_____   _____
Printed Name [N]   Printed Name [N]

_____   _____
Representing (if applicable) [O]   Representing (if applicable) [P]

_____   _____
Date   Date

Appendix A

A Preferably fill in the company that you are confidentially disclosing your invention to as the Recipient. At least put in the name of the individual here but realize that the individual often will not work alone on your project. Include the firm name in the "Representing"**O** blank under Recipient Signature. If you find yourself dealing with someone who wants to play games with who the Recipient is, you probably want to work with a different firm. Walk out!

B Signing for the idea and/or invention still gives you some protection if your invention is not patentable—but not much. At best the Recipient will believe they cannot compete with your unpatentable invention for 5 years; at worst they will hit the market just days after your first public disclosure. You will have the right to enforce the agreement for the designated number of years but you may decide that taking your next multi-million dollar idea to another company is a better idea. Don't forget to let other inventors know about your experience with a less than ethical company.

C Limiting the period to 5 years eliminates requesting a signature on an "evergreen" contract which many parties won't sign. The 5 years also means you have to get things together in that 5 years.

D Spelling out what is NOT confidential should give the recipient some assurance that you won't be badgering them over every little thing that you THINK might give a competitor a clue about what you are up to.

E You must own any "fixes" or other "improvements" that their expertise provides that become part of your patent claims, otherwise they will have you over a barrel if you require their expertise to complete your invention. **If their expertise is only needed to select among ideas you presented to them or they suggest ideas you already had and documented in your inventor's notebook or their expert suggestions are just "prior art" or "obvious," then their suggestions are not patentable and should not be included in the claims and will not require that they be a co-inventor.**

The "except"**F** clause will let you sign rights to processes to them or you may want to buy their contributions for $1 in a separate agreement. You also acknowledge that a fair award by arbitration or court action can limit your rights to their contributions. **O** You acknowledge that any patentable contributions on their part (which you will still own) will get the contributor acknowledged as a co-inventor on the patent application—otherwise any patent you get might easily be declared invalid. **H** The last sentence of the first paragraph simply makes explicit that they will have no right to infringe your patent, not even after 5 years, based on this agreement.

[I] The arbitration clause is followed by a paragraph[J] that makes it clear you won't put up with any loser hokum trying to weasel out of an arbitrator's decision that goes against them—and they don't have to put up with it from you either!

[K] I recommend you complete two copies of the agreement and each party has one copy. No matter how you do copies and originals, there is nothing that prevents tampering or removes the burden of proving your copy is correct if the originals/copies are not in agreement.

[L] I like some inclusion of what is expected of me, i.e., why are you giving me confidential information in the first place? Note that marketing-related activities require that I provide at least some of your information to others and I will not demand that those others sign non-disclosure agreements (because I already know too many would walk away and give me no marketing feedback if I demand their signature). Any "marketer" who promises full confidentiality of your idea/invention will be doing a MARKET study, not a MARKETABILITY study. While you do need to know what the market looks like, you also must find out if it is likely your invention will sell.

[M] Don't forget to get the signatures and date. Also get the printed[N] versions of names so you know whose signature you have. It is best if the Recipient sign for their firm as its representative. The fact that they do sign will be interpreted by an arbitrator as indicative that they and you relied on the belief that they were authorized to act as agent for the recipient firm. You can represent "self"[P] if you are sole inventor or you can represent the DBA or company you and your co-inventors or partners are operating under.

I see no need to make your non-disclosures any more complicated than this. If you get bogged down worrying about war, "acts of God," or whatever, you will probably have trouble getting anywhere.

In fact, even the above agreement can be too scary for some people to sign. You'll have to decide whether to move on or opt for a simpler one. If you are dealing with the right person or firm you may be willing to forego any written agreement and just count on logging their name and the date and what you discussed in your inventor's notebook.

B. Alternate Non-Disclosure Agreement - Sample

_____ agrees to maintain any and all
  (Recipient)

information regarding the idea and/or invention developed by

_____ *in confidence and not to use such*
  (Inventor)

information in any way without his or her express written

permission.

_____   _____
  (Inventor)   *(Recipient)*

Date: _____   Date: _____

    The above is about as simple as you can get. Remember though, all a signed non-disclosure form does is provide proof that someone represented that they would treat your idea/invention confidentially. It doesn't provide actual prevention of disclosure, only possible financial "protection," if the discloser can pay a judgement you get awarded to you, through the courts, after a disclosure.

    In the end, extreme protective paranoia is not the answer. Hard work and diligent pursuit, as evidenced by a solid paper trail and supported by a detailed inventor's notebook, is. Getting a successful product to market should give you all the profits you need to defend yourself via litigation if it should be necessary. If it won't, you are probably proceeding with the wrong product idea.

James E. White

C. Expense and Profit Sharing Agreement

The following parties agree to divide the profits, if any, from the _____
_____ invention project as noted in the
following lines and paragraphs.

Name	Percent Planned[B]	Minimum Percent[C]	Work Responsibilities/Duties[D]
_____	_____	_____	_____
_____	_____	_____	_____
_____	_____	_____	_____

    The parties agree to each assist the others in working diligently toward the completion of the invention and on its eventual marketing. If any party's effort, except[E] as caused[F] or agreed to by any of the other parties, should drop below _____ hours[G] in any _____ period of time,[H] the party whose effort is below the agreed minimum will forfeit[I] _____ of the above indicated Percent Planned, except that their percentage will not drop below the Minimum Percent, and that the forfeited percentage will be divided equally/proportionally[J] (circle one) between the other partners. Each party agrees to keep careful records of their efforts on behalf of the invention and acknowledges that the parties will work independently from time to time.

    [K]In making decisions that affect the project, each party retains 1 vote per full 1 percent of their then current division percentage. The parties recognize that, in the eyes of the law, any party may legally obligate the others to actions/decisions on which no vote was taken.

    [L]In the event of the death of one of the parties prior to completion and the initial sale of the invention, the percent of the profits, if any, to be directed to the estate of the decedent will be as follows while any division percentage above the indicated percent will be divided equally/proportionally[M] (circle one) between the other partners.

Name	Percent[N]
_____	_____
_____	_____
_____	_____

    [N]The parties agree to renegotiate, with 1 vote per party, the above death percentages on or about each of the anniversary dates of the signing of this agreement prior to the first product sale.

    [O]The parties agree, to the best of their ability, to pool their funds and split all expenses equally. In the event that one or more of the parties is unable to share the full expense load the party or parties carrying the lesser load agree that the party or parties carrying the greater load will, in the event of sales receipts, be paid back _____ times[P] their expenses above the lessor contributor's load before any profits shall be recognized for the project.

Q *The parties agree that any percentages of profits that are promised to parties outside this agreement are to be treated as incurred project expenses for purposes of this agreement. Such promises may arise due to outside financing, commission agreements for marketing, or other circumstances. The parties also recognize that they are individually responsible for payment of taxes on any profits from this project and that profits, for the purposes of taxes, may be different than profits as considered in this agreement. Taxes may be due even though no distribution of profits is made to the parties of this agreement.*

S *The parties agree that everybody that contributes <u>functional</u> ideas that are incorporated into the invention will be named as co-inventors on the patent application. Co-inventors need not be parties to this agreement and need not receive any compensation. The parties further agree that a party to this agreement that contributes by effort or finances but is not a contributor of functional ideas, will not be a co-inventor.*

T *The parties further agree to submit any controversy or claim arising out of this agreement, or any breach thereof, which cannot be mutually settled between them, to binding arbitration to be administered by the American Arbitration Association in accordance with its rules. Any and all available evidence may be presented to the arbitrator(s). The arbitration decisions regarding any and all matters will be final.*

In the event one party must resort to the courts after the other party fails to live up to an arbitration settlement, the parties agree that the judgement on the award rendered by the American Arbitration Association arbitrator(s) may be entered in any court having jurisdiction thereof, and that, should the plaintiff party prevail, the other party will, in addition to the court judgement, pay all costs incurred.

_____   _____ _____

_____   _____ _____

_____   _____ _____
Signature   *Printed Name*   *Date*

The above is a totally untested agreement. Its purpose is to give you some idea of what you need to think about BEFORE you join your efforts with other parties. However, I believe such an agreement, where it did not violate some well-intentioned law, should be quite adequate for an arbitrator to provide "justice." In the absence of some written agreement or some evidence by witnesses, the courts won't have much choice but split your project's profits equally, IF THERE IS A DISPUTE, no matter what you might have agreed on with a handshake.

For a general discussion of some of the paragraphs here, see Chapter 11 and the beginning of the Appendixes. The following are some more specific comments.

James E. White

^A The simple definition of profits is receipts minus expenses to a specific date but that is not necessarily the accounting definition. You can deal with the details later because they depend on the firm's legal structure and its accounting practices that you don't want to hassle with yet. Stick to one basic invention idea and the "project" to get it commercialized.

The work Responsibilities/Duties^D can either be spelled out by some recognizable title here or can refer to a separate attached sheet. You need to spell out some of this information to reduce the probability that one party is just signing on to coast to the profits. You also need to make it clear that for any specific period of time^H (say one month) the effort of each party should be at least some minimum^C or they will lose some profits^I.

Essentially you are enforcing diligent pursuit but you are leaving an escape route^E should there be genuinely nothing one party can do in a given period, or should one or more parties try to block^F effort by another party in order to "steal" part of their interest. You are also establishing a minimum^C below which a party's interest will not drop so that they cannot lose out entirely due to some unforseen circumstances. If the person's Planned Percent^B is 30% and they agree to a forfeit^I of 5, then, the first period they don't contribute (unless it is excused in writing by the other signers) their "division percent" will drop to 25%.

Make sure you specify whether a forfeited percentage is split^J equally or proportionally among those contributing. You must keep good records. The party that fails to do so will be at a disadvantage should any dispute arise.

^K While you want to try to democratically, by shares, vote on things that affect the project, you need to be aware that the law may override your plan. A party that keeps committing the project to things that the other parties would vote to override if they knew about them in advance is a party that should be reigned in or arbitrated out as fast as possible.

^L While nobody likes to contemplate death, it does happen. Especially before the project nears fruition, I think it is a good policy to shuffle the percents so that the decedent's estate gets something for the decedent's efforts but not necessarily control or more than should be reasonably due non-participants. If you think your partnership is worse than marriage before somebody dies, think what it would be like as a forced marriage with that "bleep bleep's" heirs. The people benefitting from the estate, and who didn't earn a dime of any future profits and won't contribute one lifted finger to the project's completion, are likely to want more than the decedent's fair share.

You MUST get some legal advice on this issue because the law on what "property" one can agree to distribute without spousal (or other) consent varies from state to state. Again decide whether the sacrificed percentage is distributed equally or proportionally[M]. Also agree to revisit the death percents[N] at least annually before sales are launched (a clause to do this just before launch might not be bad either). This allows you to do a fair distribution as you see how much each contributed. Remind all parties that they will NOT benefit from this percent—they will be dead. Their families will benefit. If necessary this issue (especially during renegotiations) can go to arbitration[T].

[O] Ideally all parties would contribute equally to a pool that expenses were paid out of. Unfortunately some parties usually will not be able to put as much money at risk as others. Those that do put more money at risk should be rewarded[P] for it in some way. I suggest that doubling (i.e., 2 times) their risk capital above what the smallest contributor contributed before declaring any distributable "profits" is equitable. If the risk capital is orders of magnitude higher, say $100,000 versus $10,000, then maybe you should triple the return on the extra risked capital. This kind of structure lets the idea person get the long-term rewards even if they have to bring in a financier at the outset.

[Q] If financing that demands a piece of the profits must be brought in from outside the agreeing parties, it should simply be treated as an expense to the agreeing parties. [R] Also be aware that the government will often see profits while you are still heavily into negative cash flows. Over the years this government wisdom(?) has ruined many struggling startup businesses.

One of the catches is that all your "startup," and perhaps R&D, expenses are NOT immediately deductible from your "income" for tax purposes. They must be amortized over many (such as 15) years. If your development period will be long and expensive, get some tax help up front. It may make the difference between success and failure.

One of the best games you might be able to play here is to start your business selling some product you buy first and try to at least break even on selling that product. Then go into the R&D for your product. Some R&D is immediately deductible and some isn't, so with a little forethought you may be able to keep from losing your shirt due to paying taxes on money you haven't even received yet. If your total startup and R&D expenses will be under about $50,000 and should be recovered in the first 6 months sales, you may not have to worry. If all that is over $500,000 and breakeven is 2 years out, it will make a huge difference to you.

¶ Be sure that everyone understands that ONLY co-inventors go on the patent application and that ALL co-inventors go on the patent application. Screwing up on such nits can result in an invalidated patent. While being first to market, if it's done right, can still give you a huge advantage, patent protection is even better. Numerous foreign companies (and even some U.S. companies) pay people in the U.S. to scout store shelves for new products that they can dramatically underprice. If they can knock out your patent in the first round because one or all of you fraudulently signed their patent oaths, you'll be in trouble.

¶ Again, I think, provided you get an honest and fair arbitrator, that arbitration is a far better course than the courts for settling disputes.

If you are an attorney reading the above agreement you'll notice that one thing I did not include was a buy/sell clause. I think they are a great idea but I think you should either have one in your formal partnership agreement if it pre-dates the above invention project agreement or wait till you do have a formal structure for your firm. **Even if you don't have a formal structure you always do have a legally binding entity form, most likely a partnership.**

In my opinion, the fairest buy/sell agreement simply gives any of the parties the right to make a specific offer to buy out the other parties. The agreement then obligates the party making the offer to execute it at the end of a specified time frame (30 or 60 days is reasonable) IF the parties the offer is made to DO NOT reverse it and buy the offering party out for the same offer (or better). It's a little more complicated if ownership is not equally split, but not much. The bottom line is, if the offering party offers a lowball (unfair) price under the assumption the other parties don't have the funds to make the purchase, the offerer is likely to get the unfair price crammed right down their throat when the parties receiving the offer procure external financing.

Another thing you might notice missing is mention of how payments for services to one of the agreement's participants is accounted for. Is it an expense or not? If one of the participants has 51% voting power can they vote themselves a (high) salary that never leaves any profits to distribute? Your spouse might not do that but remember, partners are (often) worse than spouses. Rational or not, the more money at stake, the worse it gets.

D. Agreement Addendum

This agreement is an addendum to the agreement dated _____
between the following parties:

_____   _____

_____   _____

_____

_____

_____

_____

_____

_____

_____

_____

_____

_____

_____

_____

_____   _____   _____

_____   _____   _____

_____   _____   _____

Signature   *Printed Name*   *Date*

Yeah, it's just a blank sheet of paper except that it references the date and parties to the original agreement. Write in whatever you want to agree to then have all parties sign and date it. If the parties had more than one agreement on the specified date then you need to get more explicit about which one is referred to.

James E. White

E. Marketing Consulting Certificate

Marketing Consulting Certificate

Redeemable by mailing originals only to James E. White & Assoc.

Surrender entitles the person and/or project named below to one hour of telephone consulting and/or off phone research time from James E. White & Assoc.

Name

Address

Address

City,St,Zip

Project

Phone _____ Fax

E-Mail

To get the most out of this certificate you should read and thoroughly understand the book first. Failure to do that will likely result in your getting less than the maximum benefit from this certificate. Other ways to get less than full value are to use it on a $100 evaluation which would make it worth only $50. You could also use it for any of the $2,000 service offers which would likely drop its value to about $200. And, of course you could use it to shoot-the-breeze or ask questions that are already answered in the book.

Only one certificate is redeemable per person and/or project. In other words, if you, as an individual, are working on several projects you can redeem the certificate for only 1 of your projects and you can never redeem another certificate again. If you are part of a group working on a project you can redeem it for that project only. No other member of the group can redeem a different certificate for more assistance on the project. How will I know? If you "cheat" the odds are that I won't know but you will. Expect the same kind of fair dealing for yourself as you do with others and you usually won't be disappointed.

Will one hour of consulting necessarily provide you with much? It might but it is impossible to tell in advance although I find I learn something new with every project I work on. On the other hand, using the certificate will give you a small opportunity to evaluate my services in a way that is specific to you. If I do things right that may later induce you to pay for services or recommend them to a friend.

James E. White

F. Word Processor Readability Analysis

The following descriptions of word processing program use specifically refer to commands, menu choices, buttons, etc. that are correct in the specific versions referenced. The authors of those programs are not under any obligation to have made these program versions consistent with previous versions nor are they obligated to make future versions compatible. You should read the text for examples of "what" is being done and follow that rather than cursing because the exact, specific "how" of your software version is not shown.

<u>Microsoft Word</u>

In Microsoft Word 97 you deal with grammar checking by first Choosing **Tools|Options** from the menubar then the dropdown menu. Next select the "Spelling & Grammar" tab. For entering your test text without spelling and grammar checking, be sure the "Check spelling as you type" and "Check grammar as you type" checkboxes DO NOT have checkmarks in them (click on them to get rid of the check mark if they do). If either or both the "Check…" options are grayed out and not responsive you'll need to consult your documentation and/or setup disc to find out how to install the spell checker and/or grammar checker. Also choose **Tools|AutoCorrect** then be sure the "Replace text as you type" box is not checked.

After entering your text, turn the same 2 "Spelling & Grammar" tab check boxes back on. Also, while you are there, be sure the "Check grammar with spelling" and "Show readability statistics" checkboxes in the "Grammar" section are checked. In the "AutoCorrect" settings box recheck the "Replace text as you type" checkbox. Make sure your language is set to "English (United States)" by choosing **Tools|Language|Set Language** then selecting "English (United States)" from the list.

Count your spelling errors (excluding correct words that are not in the spell checker's word list) and write down the number. Also count your grammar errors (punctuation errors ARE grammar errors as are incorrect plurals and tenses even if they were already counted as spelling errors) and write down the number.

If just looking at the errors marked on your screen doesn't give you the information you need to correct the errors, choose **Tools|Spelling and Grammar** from the menubar and dropdown menu. The Spelling and Grammar checker will run and show the words and sentences that probably need correction. The spell checker is much better at offering correct suggestions than the grammar checker.

If you are NOT able to fix errors by just looking at the markings on the screen (or recognize that the checker has erred or flagged a style issue), that should also give you a good clue about your communication skills. Whether you fix your errors by just looking at the screen markings or not you will need to run the **Tools|Spelling and Grammar** to see the readability statistics displayed.

Unfortunately you can't get the readability statistics without running through the entire document and acting on all the "suggestions" thrown at you. I suggest that, if you are happy with the document as written, you simply repeatedly click the "Ignore All" button. WARNING: After you click the "Ignore" or "Ignore All" buttons any error flagging on real or possible errors is removed. Before you are done with your document, ALWAYS choose **Tools|Options,** select the "Spelling & Grammar" tab, and click the "Recheck Document" button (then click "Yes," if necessary) to have the flagging restored.

In the Microsoft Word readability statistics the Flesch Reading Ease score is equivalent to the Sentence Complexity score in Grammatik which comes with Corel WordPerfect. The two word processors also come up with slightly different grade level and other scores for the same text but the differences cannot be deemed to be significant.

Corel WordPerfect

In Corel WordPerfect you deal with grammar checking by first choosing **Tools|Proofread|Off** from the menubar then the dropdown menu to prevent the spelling checker and Grammatik from flagging errors in your typing. Also select **Tools|QuickCorrect** then click the "QuickCorrect" tab and make sure both the "Replace words as you type" and "Correct other mis-typed words when possible" are NOT checked. If they have a checkmark just click on it to get rid of it.

After entering your text, turn the automatic error flagging on by choosing **Tools|Proofread|Grammar-As-You-Go.** Choose **Tools|Language|Settings** then set "English-US" as the language. For the most reasonable results you should also choose **Tools|Grammatik** then click the "Options" button and choose "Checking Styles" from the popup menu. Select either "Quick Check" or "Grammar-As-You-Go" in the "Checking Style" listbox. Count your spelling errors (excluding correct words that are not in the spell checker's word list) and write down the number. Also count your grammar errors (punctuation errors ARE grammar errors as are incorrect plurals and tenses even if they were already counted as spelling errors) and write down the number.

If just looking at the errors marked on your screen doesn't give you the information you need to correct the errors, choose **Tools|Spell Check** from the menubar and dropdown menu. The spelling checker will open and may automatically run and start showing the words that probably need correction. If the spelling checker does not run automatically after it opens, click the "Start" button (if there is no "Start" button, it is running automatically). When the spell checker is done and it asks if you want to close it click "No." Next click the "Grammatik" tab. If the grammar checker does not start automatically, click the "Start" button. When it is done also click "No" to prevent its closing. The spell checker is much better at offering correct suggestions than the grammar checker.

If you are NOT able to fix errors by just looking at the markings on the screen (or recognize that the checker has erred or flagged a style issue), that should also give you a good clue about your communication skills. Whether you fix your errors by just looking at the screen markings or not you will need to run **Tools|Grammatik** to see the readability statistics displayed. With the "Grammatik" tab selected you should see an "Options" button which you can click. A popup menu appears. Hover the mouse pointer over the "Analysis" option then click "Readability" on the next popup menu. After a few seconds the readability check should finish and display the results.

WARNING: After you click the "Skip" or "Skip All" buttons when running Grammatik, any error flagging on real or possible errors is removed. Before you are done with your document, ALWAYS choose **Tools|Grammatik** then click the "Options" button and choose "Checking Styles" from the popup menu then click the "Select" button and close the grammar checker to have the flagging restored.

If you want to be very careful with your grammar you might choose the "Very Strict" Checking Style when your document is nearing completion. If you have a long document you will likely be bored to tears before completing the run. In Grammatik you can also create your own checking style and make it even stricter than the "Very Strict" style but you will probably also discover that the number of non-errors presented for your review becomes tedious.

Fog Index

A couple of readers of drafts of this book thought that having a word processing program compute the readability of a text might be too difficult for some buyers of the book and suggested that a simple fog index would do. I have included instructions for doing a fog index but have not used them on the book for several reasons. Among them is that doing the work by hand is tedious and prone to errors and that the result is quite dependent on the "random" 100-word text segment. A fog index "grade," like the Flesch Reading Ease scores of the word processing software, cannot be considered precise.

There are several fog indexes, but one of the simplest is the one presented by Robert Gunning in *The Technique of Clear Writing*.

Gunning's basic procedure is as follows:

1. Randomly choose any 100-word segment of text.
2. Count the sentences in the segment including each of the beginning and end sentence fragments, if any.
3. Divide 100 by the number of sentences to get average sentence length in words.
4. Count all words in the segment with 3 or more syllables.
5. Add the numbers of steps 3 and 4.
6. Multiply the sum by 0.4.
7. Round the result to the nearest whole number. This is the approximate U.S. grade level required for understanding.

G. Foreign Economic Market Information

How do you get a handle on foreign economies and markets. Well, first you ask "Who in the U.S. watches such things?" The answer is the CIA. How do you pry information out of the CIA? You go to their web site and poke around. The site is very good when it isn't trashed by some ethics-stunted imbecile. The CIA site is www.cia.gov just like you'd expect.

If you just want tables of numbers, first select "Publications" then "Directorate of Intelligence Publications" then "Reference Aids" and finally *Handbook of International Economic Statistics for 1998* which currently resides at www.odci.gov/cia/di/products/hies/index.html. By clicking the "Tables" choice you can download an Excel spreadsheet with a lot of economic numbers for a lot of countries. I created my own sheet within their spreadsheet and used the GDP and Population columns in two of their sheets in combination with sorts and my own formulas to create the following table.

International
Real Gross Domestic Product Ranked by Economic Strength

	1997 US$ (Billions)	Per Capita Income	Population (Millions)	Cumulative Pop. (mlns)	Economic Strength	World GDP%
United States	$8,163.00	$30,464	267.955	267.955	103.65	21%
Japan	3,134	24,935	125.689	393.644	32.57	8%
Germany	1,840	22,419	82.072	475.716	17.19	5%
France	1,369	23,358	58.609	534.325	13.33	4%
United Kingdom	1,259	21,861	57.592	591.917	11.47	3%
Italy	1,246	21,925	56.831	648.748	11.39	3%
China	4,520	3,686	1226.275	1875.023	6.94	12%
Canada	707	23,305	30.337	1905.360	6.87	2%
Spain	645	16,493	39.108	1944.468	4.43	2%
South Korea	631	13,733	45.949	1990.417	3.61	2%
Australia	394	21,368	18.439	2008.856	3.51	1%
Netherlands	355	22,684	15.650	2024.506	3.36	1%
Brazil	1,007	6,006	167.661	2192.167	2.52	3%
Belgium	246	24,201	10.165	2202.332	2.48	1%
Switzerland	200	27,624	7.240	2209.572	2.30	1%
Mexico	726	7,499	96.807	2306.379	2.27	2%
Austria	193	23,730	8.133	2314.512	1.91	0%
Hong Kong	170	25,966	6.547	2321.059	1.84	0%
Taiwan	308	14,192	21.703	2342.762	1.82	1%
Thailand	503	8,461	59.451	2402.213	1.77	1%
Indonesia	936	4,462	209.774	2611.987	1.74	2%
Sweden	182	20,530	8.865	2620.852	1.56	0%
Argentina	353	9,861	35.798	2656.650	1.45	1%
Russia	675	4,582	147.306	2803.956	1.29	2%
Norway	112	25,455	4.400	2808.356	1.19	0%
Turkey	425	6,690	63.528	2871.884	1.19	1%

International
Real Gross Domestic Product Ranked by Economic Strength

	1997 US$ (Billions)	Per Capita Income	Population (Millions)	Cumulative Pop. (milns)	Economic Strength	World GDP%
India	1,575	1,629	967.118	3839.002	1.07	4%
Singapore	93	27,027	3.441	3842.443	1.05	0%
Malaysia	222	10,834	20.491	3862.934	1.00	1%
Iran	382	5,656	67.540	3930.474	0.90	1%
Saudi Arabia	207	10,305	20.088	3950.562	0.89	1%
Chile	167	11,511	14.508	3965.070	0.80	0%
Israel	100	18,067	5.535	3970.605	0.75	0%
Colombia	253	6,684	37.852	4008.457	0.70	1%
Poland	255	6,604	38.615	4047.072	0.70	1%
Venezuela	190	8,484	22.396	4069.468	0.67	0%
Czech Republic	120	11,653	10.298	4079.766	0.58	0%
South Africa	236	5,591	42.209	4121.975	0.55	1%
Egypt	284	4,381	64.824	4186.799	0.52	1%
Algeria	174	5,833	29.830	4216.629	0.42	0%
Pakistan	332	2,512	132.185	4348.814	0.35	1%
Philippines	249	3,272	76.104	4424.918	0.34	1%
Hungary	75	7,330	10.232	4435.150	0.23	0%
Peru	110	4,298	25.595	4460.745	0.20	0%
Romania	103	4,585	22.463	4483.208	0.20	0%
Bangladesh	200	1,596	125.314	4608.522	0.13	1%
Ukraine	115	2,280	50.448	4658.970	0.11	0%
Vietnam	128	1,704	75.124	4734.094	0.09	0%
Kazakstan	57	3,376	16.882	4750.976	0.08	0%
Nigeria	130	1,212	107.286	4858.262	0.07	0%
World	38,709	6,198	6245.756	4858.262	100.00	100%

From a "Where do I need to get patents for my invention?" standpoint the most important column is the one I have called "Economic Strength." For ease of use, the table is sorted in descending order on the column. As you can see, the countries commonly called the "Big Seven" all sorted to the top 8 with China squeezing in just above Canada. China, with its massive population and the gradual transformation of its economy, is going to become a major market as its per capita income increases. But for today you may want to skip China. The reasons are that getting any intellectual property rights are likely to be difficult and, if you do get them, enforcement is likely to be either impossible or prohibitively expensive. Evaluate and make a BUSINESS decision.

If you closely analyze the above table, you will discover that many countries are not listed. By the CIA's estimates the omitted countries only account for about $2.6 billion (mere pocket change) in GDP generated by about 1.4 billion people, i.e., $1.86 per year per person. My "Economic

Strength" column is a somewhat hokey calculation but it achieves what I think is a realistic economic market measure. The formula for "Economic Strength" is:

$$\frac{\left(\begin{array}{c}\text{Country}\\\text{Per Capita}\\\text{Income}\end{array}\right)^2 \times \text{Country Population}}{\left(\begin{array}{c}\text{World}\\\text{Per Capita}\\\text{Income}\end{array}\right)^2 \times \text{World Population}} \times 100$$

The formula provides a strong bias toward economies where people can afford to buy things. The result shows that the U.S. market is about three times as strong as the next closest economy, Japan.

The table also shows that, despite high per capita incomes, countries with relatively small populations (such as Switzerland and Singapore) will not be the place that makes or breaks your profits. If your math shows that they are, you already know that my opinion is you should kill off your idea before spending any more on it. It's not that they can't or won't generate profits for you, it's that the overhead of going into them will significantly reduce your profit margins there. On the positive side, the small size of those markets will also discourage knockoffs being created within those markets. A strong trademark and an experienced international distribution partner could easily make better business sense than an additional, expensive, patent.

Assuming my jewelry invention gets to market, I think I'll stick to the Big Seven for patent protection. I may, however, try to get Trademark protection in many of the others. Presuming I have a good trademark, that protection will probably provide sufficient discouragement to copycatters. This is particularly true since their markets may not be large enough to warrant local-production-only and I will be able to exclude them from the big markets.

To get more information about country economies than just the numbers, you can use the CIA World Factbook which can be found at www.odci.gov/cia/publications/factbook/index.html or by selecting it from the CIA's "Publications" page. You'll get information on the country's manufacturing, agricultural, and mining needs and contributions as well as more specific breakdowns of the population by gender and generalities about their military. I strongly recommend looking at the pages for the countries you want to go into just because it will give you a better sense of who those countries are. An additional online worldwide information resource is Infoplease.com available at infoplease.com or www.infoplease.com or via its sister location at infoplease.lycos.com/world.html.

For basic European market statistics with some industry and product breakdowns similar to what you would find in the U.S., you can use *European Marketing Data and Statistics* which is an annual publication by Euromonitor Plc of London, England. The business sections of larger or academic business libraries often have a copy. Except in rare, rapidly moving fields, even a 3 or 4 year old copy of this reference will be worthwhile. Despite wild stock market gyrations, the underlying economic phenomena are unlikely to have changed enough in that time frame to make the difference between your success and failure in a particular country. If you think it does you are probably optimistically plunging toward failure.

You may also be worried about cheap foreign imports from countries that you don't get a patent in and therefore where you have no legal right to stop their production. Patent protection in the U.S. gives you the right to block those knockoffs from being imported to the U.S. The U.S. Customs Service (www.customs.ustreas.gov) is there to help.

In 1997, Customs Service officers across the country confiscated $54,134,392 worth of counterfeit goods in 1,943 seizures. Over the past five years, 10,542 Customs Service seizures nationwide resulted in the recovery of $230,516,410 of phony goods. To put those numbers in perspective, in 1997 the seizures amounted to the equivalent of .00066% of the U.S. economy. In other words, a gnat's ass.

Also, in 1997, the Canine Enforcement Program recorded over 9,200 narcotic seizures resulting in the confiscation of 233 tons of narcotics with a street value of over $3.1 billion, or the equivalent of .038% of the U.S. economy. The 1998 numbers for the most common knockoff products (primarily coming from China, Taiwan, Hong Kong, Korea, and India) are shown in the following table found at www.customs.ustreas.gov/enforcem/enforcem.htm: under "Fraud Investigations" "Intellectual Property Rights (see statistics)."

James E. White

Top Goods Seized by U.S. Customs in 1998

Data represents the domestic value in U.S. dollars for FY98

Media (movies, software, music)	$ 21,773,618	29%
Computers/Parts	$ 12,620,089	17%
Wearing Apparel	$ 6,615,939	9%
Toys and Video Game Cartridges	$ 4,812,372	6%
Fans	$ 3,522,204	5%
Watches/Parts	$ 3,387,146	4%
Power Chargers/Converters/Adapters	$ 2,253,646	3%
Perfumes and Makeup	$ 2,028,445	3%
Integrated Circuits	$ 1,735,803	2%
Headwear (hats, caps)	$ 1,554,415	2%
Other Commodities	$ 15,592,827	21%
Total FY 98 Domestic Value Seized	**$ 75,896,505**	
Total Number of Seizures	**3,409**	

The point of all the above numbers is to show you that basically your fears of being ripped off by foreign knockoff artists are highly exaggerated. Yes, it does happen, and it might happen to you, but most of what you hear about is just carping. <u>Probably carping by people that had no protectable uniqueness to their products in the first place and who just flat out got out-competed in the marketplace in the second place.</u> In the unlikely event you do have a protected invention and do get knocked off and letters to the offender have no effect, I suggest you fight back first with numerous, careful (TOTALLY factual) press releases—and the assistance of the U.S. Government. An experienced attorney can provide invaluable assistance.

If you are truly enough of a smashing success to be the target of rip-off artists you should consider yourself very lucky. You should also easily have the funds to work at protecting yourself.

While you might think that the Customs Service is where you would go to get knockoffs blocked, you would be wrong. Per the Customs site:

"Customs has no authority to prevent the importation of goods which violate a PATENT *unless* [emphasis added] directed to do so by an exclusion order issued by the U.S. International Trade Commission (ITC) under the provisions of section 337 of the Tariff Act of 1930, as amended. An exclusion order by the ITC directs the Secretary of the Treasury to deny entry to imports in violation of the order. The Customs Service acts for Treasury in enforcing these orders."

It is your responsibility, as the owner of the intellectual property, to discover the violation of your rights, to try to stop it yourself with one or more cease and desist letters, and to file a complaint with the ITC.

Details on the procedures to be followed in obtaining an exclusion order can be obtained from U.S. International Trade Commission, Washington, D.C. 20436. At the time of this writing the details and forms were not on their site at www.usitc.gov but I have requested they put them there. If they are not there by the time you need to find them I suggested you also request they put them there.

What is found at their site under www.usitc.gov/reports.htm is "*Section 337 Investigations at the U.S. International Trade Commission: Answers to Frequently Asked Questions*." That gets you a nice little pdf file booklet that does a good job of describing the process and, more particularly, tells you where small business entities can get government assistance with the process. You will still have to send a letter to get complete information. Things will take time, and even with the small business assistance program, are likely to cost you some money.

The ITC can direct Customs to seize imports from repetitive violators of an exclusion order. The ITC can also issue exclusion orders against goods imported by the use of many other unfair trade practices, such as violation of TRADEMARK, COPYRIGHT, and MASK WORK registrations (and the violation of TRADE SECRETS, which are not otherwise protected by Federal law).

H. How to Make More Ideas

Are you a "hobby" inventor, a one-idea-wonder, a perpetual wannabe? You will be if you stick with only your one or two non-commercially viable ideas. But you don't have to be. You can make more ideas then weed them out till you just have ones left you believe will make commercially viable inventions from. From those inventions you will need to get even more selective in choosing the ones you finally make into products. Do it once or twice with simple products and you will soon be ready to tackle bigger or more complex projects if you want to. But first, how do you generate more ideas?

Write it down. Every time you have a "problem" think about it carefully and decide if you should make a note of it. In one of the inventor clubs I go to there is an 80 year old gentleman that makes sure he carries a piece of paper and a pencil everywhere he goes. His paper gets pretty tattered and folded before it is replaced by a new piece but he is able to keep, in one place, all of his ideas.

Does he take them all to market? No. But he does select a few and move them along sometimes stopping because of existing solutions, sometimes conflicting patents, sometimes difficult or impossible technical issues, and sometimes because some testing proves it won't sell. And yes, he does "lose" some money along the way on non-commercial ideas but the commercial successes far outweigh the losses. But he never loses the cost of a patent because he does a Provisional Application for Patent to protect his invention first then determines as quickly and as relatively inexpensively as possible the answer to the question "Will it sell?" He also has a lot of fun along the way. Sure, he has an ego—but financial success stokes it far more than inept flailing.

Is he a genius? He might be but he doesn't think so and you certainly wouldn't guess it to talk to him. What is his secret? It's two fold. First he has lots of ideas and weeds vigorously. Second is he follows through himself. He doesn't tell someone he had an idea then wait for something to happen. He makes it happen. Does he tell himself how stupid he is if an idea doesn't seem commercially viable? Not a chance. He tells himself how smart he was to determine the commercial viability before wasting a lot of money and without getting his ego all tangled up in it.

A book I highly recommend for techniques (and wisdom) on idea creation and follow-through is *The Universal Traveler: a Soft-Systems Guide to:*

Creativity, Problem-Solving, and the Process of Reaching Goals by Don Koberg and Jim Bagnall first published in 1972 and most recently in 1992. The following are some of the "gems" from the 1974 edition of that book:

"Don't fall in love with an idea. There are many of them; they are expendable." (pg. 24)

"Inventions are easy. It's the job of making them work and getting them into use which is so hard." (pg. 24)

"Avoid the trap of prejudice. Prejudice means pre-judging and that means knowing what the answer is before you begin. Try to remember how to Defer Judgement." (pg.24)

"A hard-won idea can seem more precious than it really is because it took great effort to achieve." (pg. 32)

"When a problem situation arises <u>Don't do the natural thing and ask 'What can I do about it?'</u> [...] <u>Instead ask... 'What is the real problem in this situation?'</u>" (pg. 33)

"Until you acquire idea-producing methods of your own, ideas are painful to find. Afterwards, discovering ideas is the most fun-filled part of problem solving." (pg.70)

Everybody deals with little problems everyday but few ever do anything about them. Gravity has been plaguing people for years and there are thousands of inventions that help us deal with it—you could come up with the next one. Brown sugar has been lumping for years—will you invent a once-and-for-all acceptable solution?* You can see the price of lighting your home is too high when you get your next electricity bill—but will you invent the replacement for the light bulb? Etc.

Have fun but limit your expenses to probable commercially viable solutions.

---

*Please don't send me any solutions, get a viable product on the market.

James E. White

Afterword

This book does not cover everything or every possible situation you might encounter. It is intended only as a sound starting point. By now you should see that there are many ways to "protect" an invention (but not necessarily its IDEA). The book, *Stand Alone, Inventor!*, provides specific examples of products that are protected by patents, by patents and trademarks, by patents, trademarks, and copyrights, by copyrights and trademarks, and even, apparently, by nothing at all—yet they are all profitable because they fill a need.

As much as you or I might like our IDEAS, it is not the ideas that are important, what is important is that:

1) the products resulting from the ideas have <u>sufficiently more value to the buying customer</u> than our cost of producing and getting them to the customer—otherwise we can't make a profit, AND

2) the products have value, to the customer, <u>that is sufficient to induce</u> the customer to buy them rather than holding on to cash or purchasing other products.

A product that meets those two criteria provides a positive answer to the question "Will it sell profitably?" which should be why we're in the game in the first place.

Lest you think that this book is too harsh in encouraging you to throw away ideas, look at the following chart. This chart uses nice round numbers but it is based on the experience of one product development manager at a large corporation. Using the best methods available to them for researching and discarding ideas, they launch 2 out 64 ideas and only 1 of those 2 is successful.

Estimated Cost for Launching One Successful New Product

Stage	# of Ideas	Cost Each	Cumulative Total	Launch All Ideas
Idea Screening	64	$1,000	$64,000	$64,000
Concept Test	16	$20,000	$384,000	$1,344,000
Development	8	$200,000	$1,984,000	$14,144,000
Test Market	4	$500,000	$3,984,000	$46,144,000
National Launch	2	$5,000,000	$13,984,000	$366,144,000

A successful product launch for them means that the one successful product not only has to recover its own development and testing costs ($721,000) and its own national launch costs ($5,000,000) but the launch costs of the product that failed ($5,000,000) and the abandoned development and testing costs ($3,263,000) of the 62 products that were never launched.

Since your own labor is free, obviously your costs will be slightly lower. On the bright side, the manager figured that smart management decisions prevented them from completing national launches on 62 products, 63 of which nearly certainly would have failed (counting the launched product that failed).

The "profit" margin of the 1 lonely success would have had to overcome $366,144,000 in total development and launch costs if all the initial product IDEAS were run all the way through national launch. There are a few products on the market today that could do that. With a little luck, your invention just might be the next one of them.

On the other hand, your invention, or your next invention after it, might not. While it may be hard to abandon invention ideas, it is far easier and cheaper to come up with new ideas than to try (and fail) to make a commercially unacceptable invention successful.

Did anything in this book make you mad? Did some parts cause your blood to boil a little? Are you now suffering from elevated blood pressure? If so you are seeing the classic symptoms of "fight or flight" syndrome. It is biologically built into each of us. While you can't get rid of it you must learn to control it.

As I mentioned before, separate yourself from your idea. Take pot shots at your idea and join others in taking pot shots at it. Learn from the experience and if, with careful (often inexpensive) analysis as suggested in this book, the idea lacks acceptable commercial merit, do what Thomas Edison did—generate lots more ideas to pursue. In the long run your goal should be to die comfortable—and with your name in the history books—not to bloody make people accept all of your ideas as fantastic.

Is this a good advertisement for my services? Does it look like I will provide what the customer wants? Would a "FREE" invention evaluation service better provide what the customer wants? YOU, the marketplace for invention marketing services, are the final arbiter of all the answers. I will note, however, that when selling a product, as I hope you will be doing soon, you typically just accept the money from any buyer. When selling a service, such as I or your prospective subcontract-manufacturers do, that is often not the case. Trouble clients, those that keep wanting more without contributing to profits, must be weeded out before they get a toehold. Hopefully by now

James E. White

you understand your risks and your professional product developer responsibilities well enough not to be among those weeded out.

I'll take my chances with my audience, you the buyer, just like you will with your product's audience. No editors to dilute my words, no licensee to muck up the marketing, no "net" if my expenses exceed my receipts. I will, however, keep moving forward on my next invention project, maybe get started on the "marketing" follow-on book to this one, and keep progressing one step at a time as I have over the past 10 years of self-employment.

You can do the same—just accept the risks rationally and carefully. If you are absolutely certain you have a successful IDEA and you haven't even started STEP 0 yet, I'm nearly certain you will fail. After all, how many of last year's 243,062 patent applicants were wrong? And they've done far more than you.

One of the real kickers that got me going on this book occurred at the January 1999 meeting of the local inventors club. The first inventor to talk about his work specifically said, "let's hear about what everybody's been doing this past month." The importance of this point was apparently missed by most people at the meeting who could have correctly reported "nothing new to report." I set mid-May as the completion schedule for this book. I missed that by a few months but I did keep taking the steps necessary to move the book along. Amazingly enough, it did not write itself and it certainly didn't do any of its own research.

While the tone of this book is frequently negative, I hope you now understand that is meant in positive way. While I don't want to be the "dirty rotten scoundrel" that dashes your dreams I hope you understand that the majority individual inventor patent holders just have one patent in their repertoire. After that patent failed to pay for itself they gave up or had to quit. Remember the headline "They Laughed When I Sat Down at the Piano..."? Will they laugh when you show them your patent? Not likely if you show it in the mansion you bought after you learned how to correctly play the patent game (with, I hope, a little help from the instructions in this book).

I hope my analysis of the market for this kind of book is not wrong. I am also aware that while the market need may be there, this book may not properly fill it. I have both test marketed drafts of the book and have given them away for comment. The feedback has come from all over the map and includes both extremes, "The book is a goldmine" and "It's too negative and needs an editor" [it has now been edited]. I take full responsibility for the way I modified it in response to comments.

The most heavily weighted comments were the ones I got from fellow inventors WHO HAVE (or have had) SUCCESSFUL PRODUCTS on the market. My favorite comment was from a millionaire inventor (with NO patents, by the way) who noted "Jim, that's great advice—but don't be too disappointed if people don't pay attention to it." Another telling comment came from a fellow inventor who has products on the market—but hasn't had sufficient sales to generate profits "I understand where you're coming from—but I want to try it my way first."

Now that you've read (and absorbed) it all, is the following statement true or false?

One typical difference between financially successful and unsuccessful inventors is that successful inventors have rejected a lot more ideas than their unsuccessful counterparts.

Verify the Top Characteristic of Successful Inventors

Your final homework assignment is to verify your answer. You can do that by attending the next nearby inventors club meeting (many are listed at www.inventorfraud.com/inventorgroups.htm or www.patentcafe.com/inventor_orgs) and asking questions. During your research you will very likely encounter the unsuccessful, million-idea inventor. There will be at least one in every club of 10 or more. You'll easily recognize them because they keep telling you about their ideas but never seem to answer even the most pointed questions about sales. They may want you to believe they are being financially discreet—but the probabilities are they never had an idea they rejected and very few of their ideas (usually none) have actually been completed to market.

In my opinion, success in the marketplace is the only place where the invention idea counts. Anything else is wannabe. Good luck verifying your answer and good luck getting your own financially successful invention to market.

James E. White

Index

This index is actually part traditional index and part concordance and represents a compromise between human and machine indexing. Ideally I'd like to give you a full Key Word In Context (KWIC) index, but that would make the index many times longer than the book. I get irritated at book indexes when I remember a key word or phrase from a story or point but can't find it in the index because the indexer used a different word to reference it in the index (if it is referenced at all).

This index uses primarily key words, word pairs, and three word groups as found in the text by a computer program. It also uses some specific indexing I added (such as indexing SuperClip under both SuperClip and Giant Paperclip). The general rule is for you to locate the most explicit word group that interests you then look up those page numbers, if that doesn't locate what you want or suggest a more appropriate word group to look for, back off to the next most explicit word group and look up referenced pages again (except skipping the pages you already looked at), etc.

For example if your interest is "Expired Patents," look up that phrase since prefix adjectives that precede the keyword will be indexed by the adjective (I didn't teach the computer to do "Patents, Expired"). If that fails to locate what you want look under "P" and you'll find "Patent Agent" to "Patents Make Money," pick a phrase to look for etc.

You may note that some "important" words (such as "Patent" by itself) are not in the index because they occur far too often to use as index terms. ("Patent" occurs 625 times, more than once per page.) Noise words (such as "the," 5575 occurrences) are not in the index at all and were not permitted in phrases.

James E. White

James E. White

James E. White

James E. White

James E. White

James E. White

James E. White

James E. White

James E. White

James E. White

James E. White

James E. White

James E. White

Did You Notice the Cover?

A word or two about the color of the cover. You may have noticed the color is fairly LOUD. That is not an accident. I want this book to be visible so that you can find it and refer to it as often as necessary as you go through the steps to determine the <u>profit potential</u> of your invention. I also hope that the cover will bring the book to the attention of one or two of your friends who might also have ideas <u>with profit potential</u>—and who might wish to order a copy of their own to benefit from.

While I can appreciate the general tendency of inventors to keep costs down, I will not appreciate it if you copy this book for your own use or to give away (or worse, sell). If you are tempted to do that please stop and think very carefully about your reasoning. When your invention is put on the market, will you be flattered or incensed if every prospective purchaser STEALS one? I suspect you would be incensed. Yet I know of inventors who not only STEAL intellectual property protected by the Constitution (and the law) of the United States—they BRAG about it.

In particular, software piracy (theft) seems to be not uncommon among a subset of inventors. Yes, $2000 plus is a lot to spend for a high end CAD package—but you also can get very good ones <u>that will meet your needs</u> for well under $150. Please, make it a practice to do the right and honorable thing—pay for your acquisitions just like you expect your customers to pay for your work. Why should you be secretive about your invention until you get <u>your</u> intellectual property protection (a patent) if you really don't mind if your, or anyone else's, intellectual property rights are violated?

With that intellectual property theft diatribe aside, the color of the cover was also chosen for another reason. That is to illustrate the point that what you SEE can help you. If you fail to SEE the point in much of this book, or the stories therein, my belief is that you will get many (avoidable) bumps and bruises along the way and will be unlikely to become a <u>financially successful</u> inventor. The only EYES that count when SEEING your invention are not yours, but the eyes of the BUYER who must give up hard-earned money to acquire your product.

Maybe too, the LOUD cover will make it seem like my shouting elsewhere isn't so bad.